SpringerBriefs in Computer Science

SpringerBriefs present concise summaries of cutting-edge research and practical applications across a wide spectrum of fields. Featuring compact volumes of 50 to 125 pages, the series covers a range of content from professional to academic.

Typical topics might include:

- A timely report of state-of-the art analytical techniques
- A bridge between new research results, as published in journal articles, and a contextual literature review
- A snapshot of a hot or emerging topic
- An in-depth case study or clinical example
- A presentation of core concepts that students must understand in order to make independent contributions

Briefs allow authors to present their ideas and readers to absorb them with minimal time investment. Briefs will be published as part of Springer's eBook collection, with millions of users worldwide. In addition, Briefs will be available for individual print and electronic purchase. Briefs are characterized by fast, global electronic dissemination, standard publishing contracts, easy-to-use manuscript preparation and formatting guidelines, and expedited production schedules. We aim for publication 8–12 weeks after acceptance. Both solicited and unsolicited manuscripts are considered for publication in this series.

**Indexing: This series is indexed in Scopus, Ei-Compendex, and zbMATH **

Thomas Guyet • Philippe Besnard

Chronicles: Formalization of a Temporal Model

 Springer

Thomas Guyet (iD)
Inria Grenoble Centre de recherche
Villeurbanne, France

Philippe Besnard
Rennes, France

ISSN 2191-5768   ISSN 2191-5776   (electronic)
SpringerBriefs in Computer Science
ISBN 978-3-031-33692-8   ISBN 978-3-031-33693-5   (eBook)
https://doi.org/10.1007/978-3-031-33693-5

This Springer imprint is published by the registered company Springer Nature Switzerland AG
The registered company address is: Gewerbestrasse 11, 6330 Cham, Switzerland

A chronicle is very different from history proper.

Howard Nemerov, 1988

Preface

This book is intended as an introduction to what we feel is a versatile model for temporal data, initiated by the work of Ch. Dousson in the 1990's. This temporal model, that has been named *chronicle*, has in appearance a simple definition: some symbolic events and temporal constraints between them. We believe that the possibility to depict chronicles graphically makes them attractive in various applications. Figure 1 illustrates a graphical representation of a chronicle. Events are represented by the vertices and temporal constraints are represented by edges. This particular chronicle describes a usual sequence of drug deliveries to elder epileptic patients. The labels in vertices are drug codes and the temporal constraints describe the delays between events. The situation that is described by this chronicle indicates that delivery of the antiepileptic drug N02BE01 occurs 17–28 days after the first delivery of the drug B01AC06. This gives the intuition that a chronicle can capture a complex situation. Since such graphical representations are easy to understand, the model of chronicles has been used in many different fields: network monitoring [29], diagnosis failures in web services [49], medicine [24], activity recognition [21], predictive maintenance [14], etc.

Different research groups (mainly in France) investigated the use of chronicles. The two main usages are chronicle recognition and chronicle mining. In the first case, a chronicle is specified by a human, the problem is to recognize it in temporal sequences. The second case starts from a collection of temporal sequences, and the objective is to get some interesting chronicles from them. For example, the chronicle in Fig. 1 is generated through a chronicle mining process. The intervals displayed in the temporal constraints abstract the different delays that have been observed in the drug deliveries to patients. The same chronicle can be used to recognize new elder epileptic patients just by looking at the drug deliveries. Along these research works, different definitions and notations for a chronicle have been proposed. This book attempts to reconcile these different research lines by taking a new look at the notion of chronicle. Thus, our work aims at proposing the pillars for developing new methods based on chronicles.

In the 1990s, the prominent format was Temporal Constraint Networks for which the article—with exactly this title—by Richter, Meiri and Pearl is seminal.

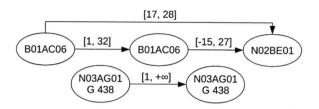

Fig. 1 Illustration of a chronicle characterizing a elder epileptic patient (borrowed from [24])

Chronicles do not conflict with temporal constraint networks, they are closely related. Not only do they share a similar graphical representation, they also have in common a notion of constraints in the timed succession of events. However, chronicles are definitely oriented towards fairly specific tasks in handling temporal data, by making explicit certain aspects of temporal data such as repetitions of an event, and by allowing a high degree of flexibility in the ordering of events. A starting point of our work was to enlighten these particularities of chronicles as well as implications for chronicle recognition and chronicle mining algorithms.

Chapter 1 is a glance at the ideas developed in the book: What are the components of a chronicle? How do they make sense together as a chronicle? How can a chronicle be (intuitively) understood and faithfully depicted? How does a chronicle relate to temporal sequences? What are chronicles used for? What does it mean for a chronicle to occur? ... Chap. 2 unfolds the formal material for chronicles, from scratch. Basic definitions and elementary results are presented in detail, with an emphasis on examples to make concrete many points mentioned in the text. Chapter 3 pursues the formal development by exploring the structure of the space of chronicles with the purpose to exhibit a structure of lattice that will form the basis of the development in the next chapters. Chapter 4 formalizes the notion of temporal sequences and how does a chronicle occur in temporal sequences. Chapter 5 investigates a more advance use case of chronicles, that is the abstraction of chronicles from temporal sequences. More specifically, we propose an original framework for extracting frequent chronicles as a significant approach to temporal data mining. Chapter 6 is the conclusion. Most of it is devoted to open questions, either about alternative options for choices made in the book or topics going beyond an introductory account of chronicles.

A list of symbols and an index are included in order to make it easier for the reader to retrieve specific items. All proofs are differed to an appendix.

Prerequisites for reading this book are just elementary notions of order theory (pre-order, lower bound, greatest lower bound, lower semi-lattice) and of universal algebra (quotient set, set closed under an operation).

This book grew out of a series of visits by the first author to the second author at IRISA in Rennes (France), during the period 2017–2021. In this respect, the help provided by Alexandre Termier, Jean-Marc Jézéquel and Bruno Arnaldi has been greatly appreciated. The authors are indebted to V. S. Subrahmanian for acting as scientific editor so that the book can be published in the Springer Briefs series.

Villeurbanne, France   Thomas Guyet
Rennes, France   Philippe Besnard
June 2022

Contents

List of Symbols

The symbols are listed here in the order in which they first appear in the book.

$\mathbb{E}$	The event types, p. 16
$\leq_{\mathbb{E}}$	Total order between event types, p. 16
$\mathbb{T}$	The temporal domain, p. 16
$\{\!\{e_1,..,e_n\}\!\}$	Multiset of event types, ordered by $\leq_{\mathbb{E}}$ (the order on event types), p. 16
$\mathscr{C} = (\mathcal{E}, \mathcal{T})$	Chronicle ($\mathcal{E}$ its multiset and $\mathcal{T}$ its set of temporal constraints), p. 16
$\mathcal{C}$	Set of chronicles, p. 16
$[l : u]$	Time interval, p. 17
μ	Multiplicity function for a multiset, p. 18
$\mathcal{T}^*$	Closure of a set of temporal constraints $\mathcal{T}$, p. 18
$\Subset$	Submultiset relation, p. 20
θ	Multiset embedding, p. 20
$\Cap$	Intersection of two multisets, p. 22
$\trianglelefteq$	Subset relation of their time interval for two temporal constraints, p. 22
$\preceq$	Less specific than or as specific as (relation between chronicles), p. 22
$\approx$	Equivalence between chronicles, p. 23
$\widetilde{\mathcal{C}}$	Set of slim chronicles, p. 30
$\widetilde{\mathcal{T}}$	Slim version of a set of temporal constraints $\mathcal{T}$, p. 30
$\widetilde{\mathscr{C}}$	Slim chronicle issued from $\mathscr{C}$, p. 30
$\widetilde{\mathcal{C}}_f$	Set of finite slim chronicles, , p. 32
$\widetilde{\mathcal{C}}_{fP}$	Set of finite slim chronicles that conform exactly with profile P, p. 32
$\widehat{\mathcal{C}}$	Set of simple chronicles, p. 32
Θ	Set of multiset embeddings, p. 34
$\prec$	The order in which the events appear in an event sequence, p. 47
$\Subset$	Occurrence of a chronicle in some stripped sequence, p. 48
$\delta(S)$	The chronicle generated from the (stripped) sequence S, p. 60
$Str(\mathcal{D})$	The set of stripped sequences in a dataset $\mathcal{D}$, p. 64
$\Lambda_\mu(\mathscr{C})$	Chronicle reduct that conforms with profile μ, p. 65

$\overline{\mathscr{C}}$ Chronicle without isolated node, p. 71
$abs(\mathcal{D})$ Chronicle abstracting a dataset $\mathcal{D}$, p. 71
$supp_{\mathcal{D}}(\mathscr{C})$ Support of a chronicle in some dataset $\mathcal{D}$, p. 75

Familiar set-theoretic symbols have the usual meaning throughout: $\emptyset$ (empty set), $\in$ (element), $\subseteq$ (proper or improper subset), $\subsetneq$ (proper subset), $\setminus$ (set difference), $\cup$ (union), $\times$ (cartesian product), 2^E (powerset of a set E), $/$ (quotient set). Lastly, we write $\{x \in E \mid P(x)\}$ to denote the set of all x in E that have property P (if the context is clear, we often omit: $\in E$).

As regards functions, the symbols have their usual meaning, too: we use $^{-1}$ to denote the inverse of a function and we use $\circ$ to denote composition of two functions. Also, we write Id to denote the identity mapping: $\mathrm{Id}(x) = x$.

We write $\vee$ to mean "or" and $\wedge$ to mean "and". Moreover, we write $\Leftrightarrow$ to mean "if and only if" for which we also sometimes use the abbreviation *iff*.

We use four kinds of right-pointing arrows. First, $\Rightarrow$ is to be read "if ... then" (and $\Leftarrow$ "only if", both of them mostly appear to indicate direction of proof). Then, $\rightarrow$ is reserved to specifying domain and codomain in the definition of functions, e.g., $f : E \rightarrow T$. Also, $\mapsto$ is reserved to specifying the image of an element of the domain, e.g., $x \mapsto x + 1$. Lastly, we write $\rightsquigarrow$ for a number of usages, roughly under the reading "maps to" (possibly referring to an informal notion of a correspondence).

For n a natural number, we write $[n]$ to denote $\{1, 2, \ldots, n\}$.

We put $\square$ at the end of a self-evident mathematical statement (lemma, ...) to indicate that no proof is provided for it.

Overcrossed symbols abbreviate negations. For instance, $\mathscr{C} \not\approx \mathscr{C}'$ is to be read as: it is not the case that $\mathscr{C} \approx \mathscr{C}'$.

List of Figures

Chapter 1
Introduction

Abstract Chronicles have been studied in the context of two analysis problems for temporal sequences: recognizing situations in temporal sequences and abstracting a set of temporal sequences. In the former problem, a chronicle is the description of a situation to be recognized in a temporal sequence. The main application field is the monitoring of system behaviour. Given a set of temporal sequences, the latter problem consists in building a chronicle that generalizes them. Such a task pertains to knowledge discovery: Chronicles then amount to synthetic pieces of information that are meant to serve as an abstraction of the data.

In this chapter, we introduce the notion of a chronicle and we develop a natural interpretation for it with respect to these two problems. We offer some intuition about our choice of a representation and go into more detail on a few technical peculiarities.

Keywords Temporal models · Situation recognition · Occurrence

1.1 Why Study Chronicles?

A temporal sequence reports what events occurred and at what time. This data type can be seen as the operating traces of a dynamical process. Such traces are commonly collected by sensors, web sites, organisations, etc. It is useful information to monitor the process, to analyse its functioning or to take decisions and actions (including repair actions).

Here are some examples of such temporal sequences and possible uses:

- communication logs: communication logs are data collected from a communication between two systems (or people). Examples are internet communications (such as TCP/IP) or aircraft communication data. The analysis of temporal sequences collected from communication protocols aims at detecting anomalies to prevent from damages. TCP/IP logs are widely used for cybersecurity. The analysis of the IP networks may help to detect DDoS (Distributed Denial of

© The Author(s), under exclusive license to Springer Nature Switzerland AG 2023
T. Guyet, P. Besnard, *Chronicles: Formalization of a Temporal Model*,
SpringerBriefs in Computer Science, https://doi.org/10.1007/978-3-031-33693-5_1

Service) attacks for instance [29]. Aircraft communication data [15] are analyzed to assess compliance with expected behaviours and to detect malfunctions.

- sensors logs: in today's factories or houses, a number of machines generate logs of their functioning. This evolution is enhanced by the development of IoT (Internet of Things) [12] technology. These logs are used to detect malfunctions, efficiency loss (e.g., when a machine consumes more energy than expected) or specific situations for which operators or the environment must react. In predictive analytics, the objective is even to forecast situations in order to take actions in advance (such as trigger repairing actions before machine malfunction).
- user activity logs: each time a user uses a service, traces from its activity are recorded and may be exploited. Retailers collect the data from loyalty cards to improve sales [35], hospitals analyse care pathways to improve care efficiency [24], social networks collect people's personal information to create ads with targeted contents.

For each of these contexts, chronicles can be used to address two analysis problems on temporal sequences:

Situation recognition. In situation recognition, an analyst wonders whether a situation of interest occurs in some temporal sequences. A temporal sequence is just low-level information that turns out to be very difficult for a human, no matter how skilled, to exploit. A situation is a higher level description likely to provide more human-oriented insight into what happens with the dynamical process under analysis. The challenge of the situation recognition problem is to devise a formal description (of a situation of interest) that can be mechanically recognized within a sequence of events. A chronicle can be seen as a template describing a situation.

Temporal sequence abstraction. Temporal sequence abstraction aims at abstracting a collection of temporal sequences into a single or a collection of representations capturing common parts of these temporal sequences. It is a situation (as defined in the previous item) that describes some behaviour(s) shared by all these sequences. It can be regarded as a mining process: The abstracted situation is generated from a collection of examples.

Both tasks are about formally representing an abstract situation that may occur in temporal sequences. However, a temporal sequence that exemplifies a situation would not do as an abstraction for a collection of sequences in the case that these sequences display different temporal delays between the same pair of events. A chronicle is to represent a situation allowing for some flexibility in the temporal delays between events.

A chronicle is essentially a collection of links between events, each link ordering chronologically two events via imposing time conditions between these two events. More precisely, ordering chronologically two events means specifying which is to happen before the other and imposing time conditions means specifying a minimal delay and a maximal delay for the second event to happen after the first event occurred (as usual in the field, it is assumed that events have a null duration).

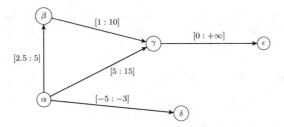

Fig. 1.1 Abstract chronicle. Event β happens in the timeframe [2.5 : 5] after α (Time units are real valued. The time is considered continuous.). Event γ happens in the timeframe [1 : 10] after β. Event γ happens in the timeframe [5 : 15] after α. Event ϵ happens after γ (or at the same time) (Do not bother about actual infinity: Please read [0 : $+\infty$] as [0 : $+\infty$].). Event δ happens in the timeframe [3 : 5] before α

Such a double condition is called a temporal constraint. Traditionally, a chronicle is represented by a graph of the kind depicted in Fig. 1.1. Each node of the graph represents an event to happen and each edge represents a temporal constraint. For instance, event γ happens from 5 to 15 time units after event α. A reason why chronicles are used in practice is their readability. The situation that is specified by a chronicle can be graphically represented so that a domain expert can easily understand it.

1.1.1 Situation Recognition

Situation recognition consists of recognizing instances of a given, complex, situation in a stream (i.e., an infinite series) of timestamped events. It is a monitoring task. A complex situation has the form of a combination of logical conditions and temporal constraints.

A situation recognizer is a combination of both a situation specification language and a computational model [39]. On the one hand, the challenge is that these situations must be specified in such a way that they can be recognized accurately (in its technical sense) as well as quickly. The specification has to be precise enough to distinguish cases of interest from dull cases. On the other hand, the recognition task must be computationally efficient (in time and in space). As applied to a stream of events, recognition is performed on the fly to avoid storing the logs. Recognition must be efficient enough for events to be processed as they arrive. Even with logs stored in large databases, efficiency of recognition is a crucial feature in practice. Thus, there is a tradeoff between expressiveness in the description of situations and computational efficiency of the recognition task.

This problem has been addressed in different research communities. They have devised different frameworks to specify complex situations and recognize them. In the formal verification community, Pnueli [60] introduces linear temporal logic (LTL) to reason about programs. Since this seminal work, numerous temporal logics have been proposed: state based logics (e.g., LTL, CTL, TCTL, Halpern-Shoham) or metric temporal logics (e.g., MTL). The metric temporal logics handle a continuous representation of time which is also our case with chronicles. We refer the reader to the book from Demri et al. [26] for an extended review of such temporal logics. These languages may be used for a situation recognition task (so called, monitoring) to specify program behaviour in an intuitive, elegant, and expressive way [72]. The computation model of these logics derives from automata theory, and more specifically from Büchi automata [34].

In the formal reasoning community, temporal action logics are used to express models of dynamical systems in a formal language [61]. This is a logical representation of the recognition problem, permitting to represent situations and to reason about them in a single formalism. The situation calculus [55] founded most of the logics representing a dynamically changing world. It defines the essential notions of action, situation and fluent, which are time-varying properties. The situation has to be understood as a static state. Actions bring changes to the world. Situation calculus is a case for modal logic. As actions in it are hypothetical, it permits to reason about possible futures of a system. Such formalisms are more commonly used for reasoning tasks, e.g. planning or model checking, that require to explore different possible futures depending on the actions.

Event Calculus (EC) [5, 46, 57, 66] is another logical formalism to reason about events and actions. While the situation calculus takes a branching time approach, EC makes the linear time assumption [73]: there is a single time line on which actual events occur. EC appears suitable for event recognition [6].

Logic-based approaches are interesting because they use a single formalism both as the situation specification language and as the formalism of the computational model. Nonetheless, they put an emphasis on the expressiveness of the situation description. For practical usages, existing tools, such as Golog [51] or DECR [58], do not seem computationally efficient. Artikis et al. proposed the RTEC system [5] which is founded on EC to recognize situation online. Based on Prolog, RTEC uses a number of techniques for increased performance and scalability.

In the constraint programming community, recognizing a complex situation is a problem of constraints satisfaction. Such a problem where variables denote event times and constraints represent the possible temporal relations between them, is then a temporal constraint satisfaction problem [65]. The main task is answering queries about scenarios that satisfy all the constraints. Dechter et al. [25] introduces the simple and general temporal constraint satisfaction problems, denoted respectively STP and TCSP, that is the verification of a set of metric temporal constraints, such as "P_1 started at least 3 hours before P_2 was terminated" on a set of timestamped events. Such a set of constraints can be represented in a constraint graph, a so called *Temporal constraint network* in the context of TCSP. Further work [13] presents a bridge between a temporal logic formalism (AL^2) and temporal constraint networks.

We must mention two research fields that address the problem of situation recognition with a more practical objective: process management and complex event processing. In these research fields, the objective is to provide situation recognition systems to support the analysis of complex dynamical systems. In process management, the analysis is essentially post-hoc whereas complex event processing is dedicated to recognize situations on-line. In these research fields, advantage is taken of the formal approaches presented above to devise situation recognition systems. For the most part, however, a computational perspective is favoured, giving up some of the complexity of situations. This is especially the case for complex event processing that may have to deal with high-speed data streams. As a consequence, in these approaches, representing situation templates is kept separate from recognizing them. On the one hand is a language to specify situations and, on the other hand is a recognition system. Chronicles have been mainly developed under this point of view in the 1990's [28, 36]. CRS (Chronicle Recognition System) [29] is a system that aims at recognizing situations specified by chronicles.

In the field of databases, situation recognition can be seen as submitting requests, in the form of temporal queries [67], to a temporal database. It is here important to distinguish two different times in the temporal database: on the one hand, the valid time of the facts in the database which is the time period during which a fact is true in the real world, and the transaction time which is the time period during which a fact is stored in the database. Only the former time is suitable for situation recognition. By the early 1990's, several temporal query languages had been proposed, and founded the TSQL2 language [68]. But many alternative temporal query languages have been proposed. Most of them are founded on temporal logics that have been introduced earlier. Their computational models are based on the database principles (efficient data structures and relational algebra). These principles enable to transform queries in order to optimize their resolution (in space and time) on dedicated data structures [50]. In the specific case of chronicle, efficient algorithms have been proposed to match chronicle on sequence databases [40].

With the emergence of new query languages in the context of the semantic Web, such as SPARQL, there are also a collection of proposals to operate SPARQL on timed data founded on temporal description logics [52]. For instance, τ-SPARQL [70], extends SPARQL with time point queries. Also, EP-Sparql [3] is a SPARQL extension for stream reasoning and complex event processing that is used in the ETALIS tool [4] or SPASEQ [38]. At the interface of semantic web and complex event processing few works were interested in chronicle [7, 14].

When process to model are complex, one can use business process models [19]. A business process model and notation (BPMN) represents a situation by a graph with different types of nodes: *events* that represent facts that occur instantaneously during process execution and *gateways* to control the divergence (logical *or*) and convergence (logical *and*) of the sequence flow. Several extensions of the original BPMN enhance the expressivity of BPMN with metric temporal constraints [18, 20, 31]. For instance, Time-BPMN is an extension of BPMN 1.2, proposed to ease the representation of temporal constraints on the occurrences of events. In [20], the authors add a constraints on the duration of the event. For each of these extensions, the authors proposed new graphical representations of the temporal constraints and algorithms that implement the recognition of such an model.

1.1.2   Abstracting Sequences

When faced with a host of temporal sequences, knowledge arises from answering the question: What do these temporal sequences (or most of them or a sample thereof or . . .) have in common? Some answers are too exceptional to focus upon because they depend on very special cases, such as: All the temporal sequences under review have the same number of events, or the delay between an event and the next event is always the same. In general, valuable answers hide deeper, in the form of a series of events of various kinds, such that each of these events is to occur within a certain time frame from the previous event in the series (which is not to be confused with the previous event in a temporal sequence). Such a series is what is called a chronicle: it captures the largest common part of the events in the series and generalizes the time frame between these events.

This problem of extracting valuable information from temporal sequence is related to the field of temporal pattern mining. In a pattern mining task, structured information—the patterns—are extracted from structured examples (eg., itemsets, sequences, graphs). Temporal pattern mining is the branch of pattern mining dedicated to extracting patterns from temporal data, such as temporal sequences. Each extracted pattern abstracts a subset of the structured examples.

Many types of patterns have been defined to analyse temporal sequences. All types of temporal pattern specify events, and it is the nature of the temporal information that differs from one type of patterns to another:

- in sequential patterns [53], the relations between every pair of events is the predecessor relationship, as events must occur in a strict order.
- in episodes [54], a predecessor relation is specified on a subset of the pairs of events. This permits to express parallel events (i.e., events that must occur in a sequence but without a predecessor relationship).
- in [44], any relationship between a pair of events is a McDermott's point relationship (before, after, meanwhile, . . .) or an Allen's temporal relationship [1] between intervals (before, after, starts, overlaps, . . .)
- in [37, 75], the events are ordered sequentially but the temporal relationship is enhanced with *time-gap constraints*. A time-gap constraint is a temporal interval with positive real values to specify the possible time frame between two successive events.
- in chronicles [27], temporal intervals are specified for some pairs of events. This combines the notion of partial order of episodes with time-gaps constraints.

The expressiveness of chronicles makes them appealing in the context of temporal pattern mining. Patterns of any kind mentioned above can be represented by chronicles. But, mining chronicles raises the issue of determining which are the chronicles of interest. The nature of temporal constraints, with temporal intervals, implies that an infinity of chronicles may be equally interesting. For computational purposes, chronicle mining algorithms make implicit or explicit choices underlying the task of determining what chronicles abstract a given set of sequences.

Which of the abstractions computed by temporal sequence mining algorithms are the most suitable? According to our intuition, a suitable abstraction is a generalisation of the sequences at hand. Considering quantitative temporal constraints, like in chronicles, there are infinitely many possible sequence generalisations, among which the least general one is a decent candidate for being a suitable abstraction of the sequences at hand. This notion of *least general generalisation* (*lgg*) invites us to consider the general framework of formal concept analysis, which formalizes *lgg* in the context of pattern mining.

Formal concept analysis (FCA) [33] is a conceptual framework to formalize pattern mining. In this framework, a pattern is obtained as a generalisation of some examples. Generalisation is achieved through an intersection operator between examples, the definition of the least general generalisation (*lgg*) relying on the assumption that the set of patterns forms a lattice. Intersection is easy to define for itemset mining (intersection of itemsets is standard set intersection) as well as more complex types of patterns (sequences, graphs) for which FCA has been extended. These are the so called pattern structures [47]. Attention must then be paid to the definition of generalisation. Balcàzar et Garriga [8] explores the FCA theory to formalize a closure system that characterizes sequential data. In this work, the author suggests that the intersection of two or more sequences is not necessarily a single sequence. Thus, the *lgg* of a set of sequences is a set of sequential patterns. Kuznetsov et al. [48] explore graphs for learning biological functions of molecules (graphs with labels). Their generalisation of graphs is based on a similarity operator on sets of graphs. Again, the generalisation of two graphs is a set of graphs. To the best of our knowledge, mining temporal sequences has only been addressed by Nica et al. [59]. Their approach is based on Relational Concept Analysis. In contrast to chronicles patterns, their patterns do not have quantitative temporal information.

To summarize, FCA is a standard formalisation of pattern mining tasks. It offers a definition of the sequence abstraction as a least general generalisation based on intersection operators. This requires to have a semi-lattice structure on the set of the patterns. It raises the question of the topological structure that can be built on the set of chronicles. Chapter 2 will show that the set of chronicles is a poset but not a semi-lattice. However, Chap. 2 will also include the description of families of chronicles that do admit a semi-lattice structure. The practical significance of such a family is discussed in Chap. 5.

1.2   The Role of a Chronicle and Its Components

Events are of various types, making matters less simple. Roughly speaking, an event type is a recurring circumstance. For example, "epileptic seizure" is an event type, it has many instances for an Epileptic patient, and any (finite) sample of these instances about the patient can be enumerated (we write sample because some instances may have eluded, e.g., the first ever, or may be disregarded for whatever reason). Thus, events in the sense of chronicles turn out to be instances of event types.

In the view of chronicles as illustrated by Fig. 1.1, each event is different. Representing an event, each node in the graph is to be given a unique label. For the label to be complete, it must disclose the event type. Then, the label is to consist of a pair (e, i), namely, the event type together with an instance number (*in a naive version*, such a number i indicates that this particular node represents the ith instance of this event type in the chronicle). Because temporal constraints thus refer to instances of an event type, a chronicle is presumably to be expressed as a set of temporal constraints specified over a chain of instances of event types.

The naive version just mentioned is the simplest format, but is rather cumbersome when handling chronicles. It is better to enumerate all instances in a single thread, the instances of the same event type being next to one another in the enumeration. The instance number in the label is the position in the enumeration. A chronicle is thus expressed as a set of temporal constraints specified over *a chain of instances* of event types, and these temporal constraints refer to instances of events type by indicating which event type and which index in the chain: see Fig. 1.2.

Notation Wherever we write X, i for some event type X and index i, the reduced fontsize (of the index) has no meaning as it serves only readability purposes.

Also, considering events as instances of event types has a consequence regarding temporal sequences. Actually, the format of a temporal sequence is not exactly a sequence of pairs (e, t) where e is an event and t a time stamp. Indeed, a sequence of timestamped events would then consist of pairs $((e, i), t)$ where e is an event type. Assuming the same event can not happen twice at the same time, the index i is otiose because it can be retrieved from the chronological order already given by the time stamps. That is, a temporal sequence can be identified with a series of pairs (event type, time stamp), each such pair reporting that an instance of event type e

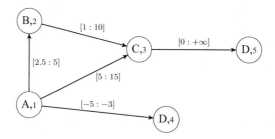

Fig. 1.2 A chronicle with events as instances of event types. Event types: A, B, C, D. Events: A, 1 (1st instance in the chronicle), B, 2 (2nd instance in the chronicle), C, 3 (3rd instance in the chronicle), D, 4 (4th instance in the chronicle), D, 5 (5th instance in the chronicle and 2nd instance of type D)

occurred at time t. Hence, writing event types e_n and time stamps t_n, a temporal sequence is simply of the form

$$(e_1, t_1), (e_2, t_2), \ldots, (e_l, t_l).$$

As a chronicle is an answer to the objective of determining a feature shared by some given temporal sequences, an appropriate chronicle must somehow show up in these temporal sequences. The idea is that a chronicle $\mathscr{C}$ can be identified with a family[1] of short sequences $\mathfrak{F}_\mathscr{C}$ (intuitively, constituting a pattern). For example, the pattern arising from the chronicle in Fig. 1.2 is D, A, B, C, D. The family then consists of *all* short sequences

$$(D, t_{D_4}), (A, t_A), (B, t_B), (C, t_C), (D, t_{D_5})$$

where each t_X has a single value (taken from $\mathbb{R}$) such that $t_{D_4} + 3 \leq t_A \leq t_{D_4} + 5$ and $t_B + 2.5 \leq t_A \leq t_B + 5$ and $t_A + 5 \leq t_C \leq t_A + 15$, etc. In a nutshell, if a chronicle $\mathscr{C}$ is a common feature for some given temporal sequences $\mathfrak{T}$, it is expected that, for each temporal sequence S in $\mathfrak{T}$, at least one sequence in $\mathfrak{F}_\mathscr{C}$ is in fact a subsequence of S.

How can the condition that a sequence in $\mathfrak{F}_\mathscr{C}$ is a subsequence of S be captured? A natural way is to check whether it is possible that all instances (of the event types) in the chronicle are mapped to a subsequence of the events in the temporal sequence under consideration. The mapping must be such that the delays indicated in the temporal constraints of the chronicle are met. In general, for the same chronicle and the same temporal sequence, more than one mapping can be found (see the dashed arrow in Fig. 1.3).

We identify such a subsequence with the notion of an *occurrence* of the chronicle, and we say that the chronicle occurs in the temporal sequence.

An order between events is required to arrange for the succession of the events in a temporal sequence. Events in a sequence are ordered by their time stamp, but in case of identical time stamps, the type of the events is used to order them. For instance, in Fig. 1.3, at time 81.8 the two events are also ordered according to event types (lexicographical order). A unique order on the event types, applicable both to chronicles and temporal sequences, seems sensible. In particular, this order is to be invariant over chronicles. Summing up, a set of event types is assumed, endowed with a total order (Sect. 1.3.1 presents some rationale for all this).

---

[1] Infinite family: in Fig. 1.2, a short sequence for each real-valued date for the moment that A, 1 happens, and similarly for B, 1 (in the interval [2.5 : 5] from A, 1, though) and so on. There can be more than one family if the delays in the chronicle are too loose to determine a unique chronological order. In such a case, simply iterate to range over these possible chronological orders.

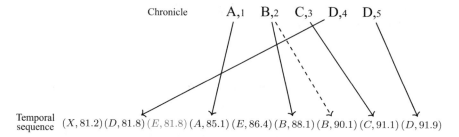

Fig. 1.3 A mapping from a chronicle (see Fig. 1.2) to a temporal sequence (plain line arrows). The dashed arrow corresponds to an alternative mapping, differing on the image of $B, 2$. The color of the third event highlight that E occurs at the same time as the second event (see text for details)

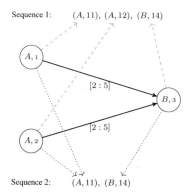

Fig. 1.4 Illustration of arbitrary mappings of a chronicle with three events: $(A, 1)$, $(A, 2)$ and $(B, 3)$, to two sequences. The (valid) mapping to the sequence on the top, in green, maps each event of the chronicle to a different event in the sequence. The (invalid) mapping to the sequence on the bottom, in red, maps the two events of type A with the same event in the sequence

1.2.1 Injective Mapping from Events in Chronicles to Events in Sequences

Figure 1.4 illustrates potential mappings of a chronicle with two different sequences. The specificity of this chronicle is to have two events with the same event type A. We point out here that the mapping of a chronicle onto a sequence must map each event of type A with different instance of this event type in the sequence.

In the example depicted in Fig. 1.4, the mapping for sequence 1 (- - ⟶ in green) is valid. However, the mapping for sequence 2 (⋯⟶ in red) is not valid. In the latter mapping, the events $(A, 1)$ and $(A, 2)$ in the chronicle both map to the event $(A, 11)$ in sequence 2, namely, an instance of event type A at time 11. Such a mapping is invalid.

This means that the mapping of a chronicle in a sequence is not simply a conjunction of individual constraints on the nature of events and their time stamps. In the example, the mapping with sequence 2 satisfy every constraints: each event of the chronicle maps an instance of the same event type and every temporal constraints are satisfied. Nonetheless, it is not considered as valid.

To sum up, no two distinct events in the chronicle, even in the case that they have the same type, are mapped to the same event in a sequence. Mathematically speaking, the mapping is injective.

Remark Injectivity of the mapping from the events of a chronicle to the events of a sequence is not encountered in alternative temporal models. For instance, in the model of Piel et al. [15], a spurious transformation of the chronicle in Fig. 1.4 is:

```
( ((A then 2) B) at most 5 ) & ( ((A then 2) B) at most 5) )
```

In contrast to chronicle recognition, the recognition algorithm for Piel's temporal model neither enforces that the two events of type A have to be mapped to distinct events in the sequence, neither enforces that the events of type B have to be mapped to the same event in the sequence.

1.2.2 Consistency and Redundancy of Temporal Constraints

Beside a list of events, the second component in a chronicle is a set of temporal constraints.

Although chronicles were first proposed [27] in 1999, there had been work on temporal constraints beforehand, most notably [25] (further work is e.g. [13]). There is a close relationship between chronicles and graphs of temporal constraints. Actually, chronicles suffer from similar problems as graphs of temporal constraints do.

More specifically, the set of temporal constraints may be inconsistent. Checking the inconsistency of a set of temporal constraints is a problem in itself [65]. It is solved in polynomial time, using the Floyd-Warshall algorithm on the distance graph representing the graph of temporal constraints [25]. In the same manner, a set of temporal constraints, if redundant, can be simplified.

In our formal development, the set of temporal constraints in a chronicle is kept as it is: We are not interested in how a chronicle may be transformed in another *equivalent* chronicle. We are not concerned with simplifying a redundant set of temporal constraints. In particular, we do not enforce chronicles to be free of redundant temporal constraints. Nonetheless, to deal with possible inconsistency in temporal constraints, we introduce the notion of well-behaved chronicles.

The next section goes into more details about the notion of temporal constraints in chronicles.

1.3   Notions of Temporal Constraints

Temporal constraints have been widely studied [65]. To prevent misunderstandings with the notion of a chronicle, this section gives some intuitions about the semantics of temporal constraints in chronicles.

1.3.1   Edge Directions and Negativity of the Temporal Constraints

We can notice in the previous examples that some temporal constraints display negative values. For example, in Fig. 1.2, there is a temporal constraint between $A, 1$ and $D, 4$ with interval $[-5 : -3]$. Intuitively, such a chronicle is to specify that that instance of D must occur between 3 and 5 time units *before* the instance of A. This choice may seem unnatural as we might prefer to deal with a temporal constraint $[3, 5]$ from $D, 4$ to $A, 1$ to specify that A must occur between 3 and 5 time units *after* the instance of D. In this case, the edge would follow the direction of the arrow of time.

The choice we make in the formal development is to direct the edges according to the order[2] on the event types (when its comes to algorithms, this is likely to be more of an advantage than a drawback [22]). Assuming that D is a greater event type than A, the edge goes from $A, 1$ to $D, 4$.

A strength in the way temporal constraints are expressed in a chronicle is that the boundaries of the interval can be negative as well as positive. For instance, the edge between $B, 2$ and $C, 3$ could be labeled with the interval $[-2 : 4]$ meaning that that instance of C must occur at least two time units before the instance of B and at most 4 time units after the instance of B. In such a case, the temporal constraint enforces neither that B must happen before C nor that C must happen before B, both possibilities are left open although not symmetrically: if B is to happen after C, it can do so at the latest 2 time units after C but if C is to happen after B, it can do so up to 4 time units after B.

Such temporal constraints are a strength of the model. It addresses the abstraction problem pointed out by Moerchen et al. [56]. Consider modelling a process characterized by the events A and B occurring at the same time. In the raw data, because of some negligible disturbances, the observed temporal sequences have the forms $(A, t_0)(B, t_0 + \varepsilon)$ and $(B, t_0)(A, t_0 + \varepsilon)$ where ε is relatively small compared to t_0. A chronicle-based model can abstract the forms of the sequences by means of a unique chronicle with the single temporal interval $[-\varepsilon, \varepsilon]$. This avoids having two different models to represent the same process behaviour.

---

[2] Remember, in the paragraph immediately prior to Sect. 1.2.1, a total order upon the event types is assumed.

Finally, the direction of an edge between two nodes sharing the same event type is determined by the index of the events, from lower to higher. As boundaries of temporal constraints can be negative, this means that the indices of events sharing the same event type need not be consistent with the chronological order in a sequence. For instance, $(A, 1)$ is to occur after $(A, 2)$ if the chronicle has a temporal constraint $[-\infty, 0]$ between them.

1.3.2 Unlimited Edges vs. Absence of Edges

Temporal constraints in chronicles allows for unlimited temporal constraints. In Figs. 1.1 and 1.2, we have temporal constraints between event types $C, 3$ and $D, 5$ that is $[0 : +\infty]$ meaning intuitively that the instance of D must occur after (or simultaneously with) the instance of C with no further condition upon the delay.

Extrapolating the case of such a neutral boundary to both ends of the interval, there could be a temporal constraint with interval $[-\infty, +\infty]$ between two events X, i and X', i'. Would it make a difference with the *absence* of any temporal constraint between X, i and X', i'?

In the formal development of chronicles, we will distinguish between (1) the case that a single temporal constraint exists between two events, with the interval of this constraint being $[-\infty, +\infty]$ and (2) the case that there is no temporal constraint at all between these two events. However, when it comes to occurrences of the two possible chronicles (depending on whether (1) or (2) is the case) in a temporal sequence, it makes no difference.

1.3.3 What Is the Meaning of Multiple Temporal Constraints Between Two Events?

In the literature, chronicles used to have at most one edge between any two events. In the formal development, we remove this limitation and allow for an unlimited number of edges between two events. Contrary to [64], all edges must be understood as a conjunction of temporal constraints on the relative time at which events must occur in a sequence.

Figure 1.5 illustrates a chronicle (on the left) with two edges between the two events. From the viewpoint of chronicle recognition, this chronicle can be simplified into the chronicle on the right with the single edge combining the two constraints. The delay between $A, 1$ and $B, 2$ must be larger than 2 according to the edge at the bottom and it must be smaller than 3 according to the edge at the top. As an aside, if the intervals were disjoint, it would not be possible to comply with the temporal constraints, hence it is assumed that such is not the case (the notion of well-behaved chronicle in the next chapter is rooted in this assumption).

Fig. 1.5 A chronicle with multiple edges between A, 1 and B, 2, reduced to a single edge chronicle

Fig. 1.6 Abstracting a set of temporal sequences: alternative choices

What reason can there be to allow for multiple temporal constraints between the same events of a chronicle? Such a reason arises in the context of temporal sequence abstraction, and more specifically in the case that there are multiple mappings of a chronicle into a sequence. Consider abstracting the following simplistic set of temporal sequences:

$$s_1 = (A, 3), (B, 4), (B, 6)$$
$$s_2 = (A, 3), (B, 5), (B, 8)$$

Figure 1.6 depicts four chronicles with single edges that can be recognized in both s_1 and s_2. Which choice is the most suitable? It is difficult to have a definite answer to this question for any context.

Warning Although we have been talking of "temporal sequence", we will speak of "event sequence" in the formal developments to insist on the fact that the elements in such a sequence are events exactly in the sense of Definition 2.1 of an event.

Chapter 2
A Formal Account of Chronicles

Abstract This chapter gives the first elements of a formal exposition of chronicles. It provides a way to express them mathematically as well as how to depict them as diagrams. Mathematically, a chronicle is a pair of a multiset of events types and a set of temporal constraints. Then the chapter starts to report our investigation about how the space of chronicles can be structured. More specifically, it introduces a relation between chronicles which is based on the notion of multiset embedding, specific to chronicles. This relation induces a preorder on the space of chronicles. This chapter gives to the reader a deep understanding of the core definition of chronicles through the use of multiple examples to illustrate the core definitions and the limitation encountered with this first definitions. In addition, as the graphical representation of chronicle is a graph, we highlight the differences between the proposed order between chronicles and the classical subgraph isomorphism relation.

Keywords Preorder · Multiset · Temporal constraints

2.1 Preliminary Definitions

This section introduces the basic definitions, in particular for events and chronicles. The notion of chronicles has been introduced in the literature of stream event analysis [36] of pattern mining [16, 21, 24]. There are multiple different mathematical formalisation of the notion but introduced for practical reasons. From our point of view, these definitions were mathematically ambiguous and can not be used for formal developments.

In our objective to provide a deep understanding of the space of chronicles, we propose our own definition of a chronicle. The reader who already encountered the notion of chronicle may be surprised by the proposed definition which defines a set of objects much larger as usual. Note that the next chapters will introduces subclasses of chronicles which matches better the classical definition.

The notion of chronicles we introduce has the advantage to clearly highlight the hidden assumptions that have been made in the previous definition of similar objects.

T. Guyet, P. Besnard, *Chronicles: Formalization of a Temporal Model*,
SpringerBriefs in Computer Science, https://doi.org/10.1007/978-3-031-33693-5_2

2.1.1 Event Types and Events

Definition 2.1 $\mathbb{E}$, the event types, is a nonempty set endowed with a total order $\leq_{\mathbb{E}}$. $\mathbb{T}$, the temporal domain, is a nonempty subset of $\mathbb{R}$ closed under substraction. An **event** (or timestamped event) is an ordered pair (e, t) such that $e \in \mathbb{E}$ and $t \in \mathbb{T}$.

Throughout the chapter, we use without further mention uppercase letters A, B, $C, \ldots$ for event types and the alphabetical order for $\leq_{\mathbb{E}}$. For brevity, we omit commas between event types, writing e.g. $AABBBDEE$ for A, A, B, B, B, D, E, E.

2.1.2 Chronicles

Definition 2.2 (Chronicle: Multiset and Temporal Constraints)   A **chronicle** is an ordered pair $\mathscr{C} = (\mathcal{E}, \mathcal{T})$ where

- $\mathcal{E}$ is a finite ordered **multiset**, i.e., $\mathcal{E}$ is of the form $\{\{e_1, \ldots, e_n\}\}$ (in which repetitions are allowed) such that

  - $e_i \in \mathbb{E}$ for $i = 1, \ldots, n$
  - $e_1 \leq_{\mathbb{E}} \cdots \leq_{\mathbb{E}} e_n$;

- $\mathcal{T}$ is a set of **temporal constraints**, i.e, expressions of the form $(e, o_e)[t^- : t^+](e', o_{e'})$ such that[1]

  - $e, e' \in \mathcal{E}$
  - $t^-, t^+ \in \overline{\mathbb{T}}$
  - $o_e, o_{e'} \in [n]$
  - $t^- \leq t^+$
  - $o_e < o_{e'}$
  - $e_{o_e} = e$
  - $e_{o_{e'}} = e'$.

We denote by $\mathcal{C}$ the set of all chronicles over $\mathbb{E}$ and $\mathbb{T}$.

Convention $(\emptyset, \emptyset)$ is considered a chronicle.

Lemma 2.1 *In a temporal constraint, $e \leq_{\mathbb{E}} e'$.*   □

The property in Lemma 2.1 does *not* cause a limitation in expressiveness since Definition 2.1 states that $\mathbb{T}$ must be closed under substraction: Considering $u > l > 0$ in $\mathbb{T}$, a temporal constraint imposing that e occurs in the delay $[l : u]$ *after* e' despite $e \leq_{\mathbb{E}} e'$ is to be written $(e, o_e)[-u : -l](e', o_{e'})$ so that it is necessary for $-u$ and $-l$ to be in $\mathbb{T}$ (which is obtained by $(u - u) - u \in \mathbb{T}$ and similarly for $-l$).

---

[1] $\overline{\mathbb{T}}$ is to denote $\mathbb{T}$ (a subset of the real numbers, see Definition 2.1) extended with two elements, $-\infty$ and $+\infty$, resp. least and greatest for $\leq$. It is convenient (see e.g. Definition 2.4, Footnote 3) to assume $t^- \neq +\infty$ and $t^+ \neq -\infty$.

Clarification In a temporal constraint $(e, o_e)[t^- : t^+](e', o_{e'})$, we take $[t^- : t^+]$ to denote the interval in $\mathbb{R}$, not in $\mathbb{T}$. What about the case that t^- is $-\infty$ and/or t^+ is $+\infty$? For the sake of a uniform notation, a temporal constraint is always written using square brackets even in the case that t^- is $-\infty$ or t^- is $-\infty$ but his does not mean that actual infinity is considered: in a temporal constraint, $[-\infty : t^+]$ is to be identified with the interval $] - \infty : t^+]$ of the real numbers (similarly, $[t^- : +\infty]$ with $[t^- : +\infty[$ and $[-\infty : +\infty]$ with the real numbers).

In the above definition of a chronicle as an ordered pair $\mathscr{C} = (\mathcal{E}, \mathcal{T})$, the multiset $\mathcal{E}$ is required to be finite but there is no such requirement for $\mathcal{T}$. Hence the next notion of a finite chronicle.

Definition 2.3 A chronicle $\mathscr{C} = (\mathcal{E}, \mathcal{T})$ is said to be finite iff $\mathcal{T}$ is finite.

Example 2.1 (Graphical Representation of Chronicles) Figure 2.1 is a graphical representation of the chronicle $(\mathcal{E}, \mathcal{T})$ where

$$\mathcal{E} = \{\!\{ABBC\}\!\}$$

$$\mathcal{T} = \left\{ \begin{array}{c} (A, 1)[1 : 5](B, 2) \\ (A, 1)[3 : 6](B, 2) \\ (A, 1)[9 : 15.5](B, 3) \\ (B, 2)[7 : +\infty](B, 3) \\ (A, 1)[-5 : -3](C, 4) \end{array} \right\} .$$

Reminder Again, please note that wherever we write E, n for some event type E and number n (as in Fig. 2.1 for example), the reduced fontsize of the number has no meaning as it serves only readability purposes.

Remark Although rather dull, the following property is to be kept in mind: For all temporal constraint $(e, o_e)[t^- : t^+](e', o_{e'})$ in a chronicle, the interval $[t^- : t^+]$ is always nonempty.

Fig. 2.1 Graphical representation of the chronicle of Example 2.1

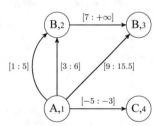

Fig. 2.2 Implicit constraints
induced in a chronicle

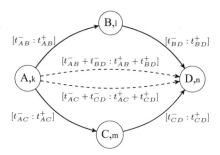

In the context of chronicles, multisets can be identified with functions $\mu : \mathbb{E} \to \mathbb{N}$.

Notation For a multiset $\mathcal{E}$, let $\mu_{\mathcal{E}}$ be the multiplicity function of $\mathcal{E}$, that is, for all $e \in \mathbb{E}$, $\mu_{\mathcal{E}}(e) = n$ where n is the number of times e occurs in $\mathcal{E}$ (n is zero when e does not occur in $\mathcal{E}$). We write $e \in \mathcal{E}$ for $\mu_{\mathcal{E}}(e) \geq 1$.

Lemma 2.2 *Let $\mathscr{C} = (\mathcal{E}, \mathcal{T})$ be a chronicle where $\mathcal{E} = \{\!\{e_1, \ldots, e_n\}\!\}$. Then, for all $(e_i, o)[t^- : t^+](e_j, p)$ in $\mathcal{T}$, it is the case that $e_o = e_i$ and $e_p = e_j$.*

Proofs in Appendix A Except for lemmas whose statement ends with the $\square$ symbol (indicating that the proof is immediate), please see the appendix for the proofs.

Taking advantage of Lemma 2.2, we can without loss of generality, assume that temporal constraints in a chronicle where $\mathcal{E} = \{\!\{e_1, \ldots, e_n\}\!\}$ are of the form [2]

$$(e_i, i)[t^- : t^+](e_j, j), \ i < j.$$

There can be a problem in the case that several temporal constraints amount to imposing different requirements to hold between two given events as follows.

▶ $(A, o)[t_{AC}^- : t_{AC}^+](C, o_C)$ together with $(C, o_C)[t_{CB}^- : t_{CB}^+](B, o')$ demand that (B, o') occurs in the interval $[t_{AC}^- + t_{CB}^- : t_{AC}^+ + t_{CB}^+]$ wrt (A, o),

▶ $(A, o)[t_{AD}^- : t_{AD}^+](D, o_D)$ together with $(D, o_D)[t_{DB}^- : t_{DB}^+](B, o')$ demand that (B, o') occurs in the interval $[t_{AD}^- + t_{DB}^- : t_{AD}^+ + t_{DB}^+]$ wrt (A, o).

This phenomenon, which is depicted in Fig. 2.2, motivates the notion of a well-behaved chronicle.

---

[2] In any case, the format $(e, o)[t^- : t^+](e', o')$ overdetermines the temporal constraint: Clearly, $o[t^- : t^+]o'$ would be enough because o (resp., o') uniquely determines e (resp., e'). For readability, however, the longer format is adopted. A further case in favour of the longer format is offered in the discussion about Example 2.3 on p. 22.

Fig. 2.3 A chronicle that
doubly fails to be
well-behaved (Example 2.2)

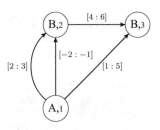

Definition 2.4 A **well-behaved chronicle** $\mathscr{C} = (\mathcal{E}, \mathcal{T})$ is such that no constraint[3] in $\mathcal{T}^*$ has $t^- > t^+$ where $\mathcal{T}^*$ can be defined as the smallest set of temporal constraints $\mathcal{X}$ such that

- $\mathcal{T} \subseteq \mathcal{X}$,
- if $(e, o_e)[t_1^- : t_1^+](e', o_{e'})$ and $(e, o_e)[t_2^- : t_2^+](e', o_{e'})$ are in $\mathcal{X}$ then
$$(e, o_e)[\max(t_1^-, t_2^-) : \min(t_1^+, t_2^+)](e', o_{e'}) \text{ is in } \mathcal{X},$$
- if $(e, o_e)[t_1^- : t_1^+](e', o_{e'})$ and $(e', o_{e'})[t_2^- : t_2^+](e'', o_{e''})$ are in $\mathcal{X}$ then
$$(e, o_e)[t_1^- + t_2^- : t_1^+ + t_2^+](e'', o_{e''}) \text{ is in } \mathcal{X}.$$

The last two clauses in Definition 2.4 give rise to two ways for a chronicle to fail to be well-behaved, see Example 2.2.

Notation For a set of temporal constraint $\mathcal{T}$, we denote by $\mathcal{T}^*$ the closure defined as the smallest set of temporal constraint introduced in Definition 2.4.

> *Example 2.2 (Two Ways for a Chronicle to Fail to Be Well-Behaved)* The chronicle depicted in Fig. 2.3 has as its multiset $\{\!\{ABB\}\!\}$. There are two reasons why this chronicle is not well-behaved.
>
> First, there are two edges from A to $(B, 2)$, one with interval $[2 : 3]$ and one with interval $[-2 : -1]$. That is, among the temporal constraints of the chronicle are $(A, 1)[2 : 3](B, 2)$ and $(A, 1)[-2 : -1](B, 2)$. By the second item for $\mathcal{T}^*$ in Definition 2.4, these two edges induce an implicit constraint $(A, 1)[\max(2, -2) : \min(3, -1)](B, 2)$, i.e., $(A, 1)[2 : -1](B, 2)$ which does exhibit the violation $t^- > t^+$.
>
> Second, there is an edge from A to $(B, 2)$ with interval $[2 : 3]$ and an edge from $(B, 2)$ to $(B, 3)$ with interval $[4 : 6]$. Due to the third item for $\mathcal{T}^*$ in Definition 2.4, these two edges induce an implicit constraint $(A, 1)[2 + 4 : 3 + 6](B, 3)$, i.e., $(A, 1)[6 : 9](B, 3)$. There is also an edge from A

(continued)

---

[3] They can fail $t^- \leq t^+$, otherwise they are temporal constraints due to the requirement that $\mathbb{T}$ must be closed under substraction hence for every $t \in \mathbb{T}$, $(t - t) - t \in \mathbb{T}$ i.e. $-t \in \mathbb{T}$ and if t_1, t_2 are in $\mathbb{T}$ then $t_1 + t_2 = t_1 - (-t_2) \in \mathbb{T}$; as to $+\infty$ and $-\infty$, min and max can be extended in an intuitive way as well as sum (assuming that $t^- \neq +\infty$ and $t^+ \neq -\infty$).

Example 2.2 (continued)
to $(B, 3)$ with interval $[1 : 5]$. Then, according to the second item for $\mathcal{T}^*$
in Definition 2.4, the implicit constraint $(A, 1)[6 : 9](B, 3)$ together with
the explicit constraint $(A, 1)[1 : 5](B, 3)$ induce a further implicit constraint
$(A, 1)[\max(6, 1) : \min(9, 5)](B, 3)$, i.e., $(A, 1)[6 : 5](B, 3)$ which exhibits
the violation $t^- > t^+$.

By its name and its nature, a temporal constraint is a limitation. It requires a
definite time frame to elapse between two given events occurring. As such a time
frame must be expressed (see Definition 2.2) using the extended real numbers,
temporal constraints of the form $(e, o_e)[-\infty : +\infty](e', o_{e'})$ seem at first sight
neutral, imposing no limitation at all. To check (see Lemma 4.3) whether such a
feeling is right, it is convenient to introduce a name for these temporal constraints.

Definition 2.5   Given a chronicle $\mathscr{C} = (\mathcal{E}, \mathcal{T})$, any $(e, o_e)[-\infty : +\infty](e', o_{e'})$ in
$\mathcal{T}$ is called an **unlimited temporal constraint**.

2.2   Multiset Embeddings

Presumably, comparing chronicles starts with comparing their multisets. A key
device is then multiset embedding, as introduced in the next definition.

Definition 2.6 (Multiset Embedding)   For $\mathcal{E} = \{\!\{e_1, \ldots, e_n\}\!\}$ and $\mathcal{E}' = \{\!\{e'_1, \ldots, e'_m\}\!\}$, we write $\mathcal{E}' \Subset \mathcal{E}$ to mean that there exists $\theta : [m] \to [n]$ such
that $e'_i = e_{\theta(i)}$ for $i = 1, \ldots, m$ and θ is strictly increasing. We call θ a **multiset
embedding**.

Fact 1   The required equality $e'_i = e_{\theta(i)}$ turns θ into a mapping from $\mathcal{E}'$ to $\mathcal{E}$ by
setting $\theta(e'_i)$ to be $e_{\theta(i)}$. $\square$

Remark   The requirement that θ be strictly increasing will turn out to be crucial
(see Sect. 3.5). The situation depicted in Fig. 2.4 is actually precluded for a multiset
embedding: The first copy of A in $\mathcal{E}' = \{\!\{AA\}\!\}$ is mapped to the 6th copy of A in
$\mathcal{E} = \{\!\{AAAAAAA\}\!\}$ while the second copy of A in $\mathcal{E}'$ is mapped to the 4th copy of
A in $\mathcal{E}$ (i.e., $6 = \theta(1) > \theta(2) = 4$).

Actually, there need not be a unique multiset embedding between two multisets.
The underlying reason is that $\Subset$ being by Definition 2.6 equivalent to the existence
of a multiset embedding, $\Subset$ amounts to a notion of a submultiset (for various notions
of a submultiset see [10]), as shown by the following lemma.

Lemma 2.3 $\mathcal{E}' \Subset \mathcal{E}$ *iff for all* $e \in \mathbb{E}$, $\mu_{\mathcal{E}'}(e) \leq \mu_{\mathcal{E}}(e)$.

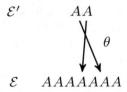

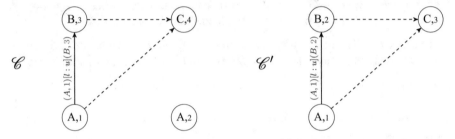

Fig. 2.4 Multiset embeddings must be order-preserving. θ at right fails to be a multiset embedding because the relative order between the copies of A in $\mathcal{E}'$ is not preserved when θ maps them to $\mathcal{E}$

Fig. 2.5 Independence of $(\mathcal{E}, \mathcal{T}'')$ or $(\mathcal{E}', \mathcal{T}'')$ being a chronicle regardless of $\mathcal{E}' \Subset \mathcal{E}$ (Example 2.3)

It is important to notice that, for $\mathcal{E}' \Subset \mathcal{E}$, neither $(\mathcal{E}, \mathcal{T}'')$ is a chronicle entails $(\mathcal{E}', \mathcal{T}'')$ is a chronicle nor its converse hold. The following example illustrates this remark.

Example 2.3 ($(\mathcal{E}', \mathcal{T})$ can be a chronicle while $(\mathcal{E}, \mathcal{T})$ is not—even if $\mathcal{E}' \Subset \mathcal{E}$)
Consider $\mathcal{E} = \{\!\{AABC\}\!\}$ and $\mathcal{E}' = \{\!\{ABC\}\!\}$. First, assume $(\mathcal{E}', \mathcal{T}'')$ to be a chronicle and take $\mathcal{T}''$ to include the temporal constraint $(A, 1)[l : u](B, 2)$.[4] For $(\mathcal{E}, \mathcal{T}'')$ to be a chronicle, it must be the case that $e_p = e'$ for every temporal constraint $(e, o)[l : u](e', p)$ in $\mathcal{T}''$. However, $(A, 1)[l : u](B, 2)$ is such that $e_2 \neq B$ because e_2 (the second item in $AABC$) is A. Second, assume that $(\mathcal{E}, \mathcal{T}'')$ is a chronicle. Take $\mathcal{T}''$ to include the temporal constraint $(A, 1)[l : u](B, 3)$. For $(\mathcal{E}', \mathcal{T}'')$ to be a chronicle, it must be the case that $e'_p = e'$ for every temporal constraint $(e, o)[l : u](e', p)$ in $\mathcal{T}''$. However, $(A, 2)[l : u](B, 3)$ is such that $e'_3 \neq B$ because e'_3 is C. All this is depicted in Fig. 2.5.

---

[4]In this example, we prefer the use of $\mathcal{T}''$ instead of $\mathcal{T}$ to state that the set of constraint is not more specific to a multiset than the other.

Interestingly, if temporal constraints were of the form $o[l : u]p$ as mentioned in Footnote 2 then it would indeed be the case that, given a chronicle $\mathscr{C} = (\mathcal{E}, \mathcal{T})$, if $\mathcal{E} \in \mathcal{E}'$, $\mathscr{C}' = (\mathcal{E}', \mathcal{T})$ would also be a chronicle. The downside is that the same temporal constraint would not need to have the same meaning in the two chronicles! This is illustrated by Example 2.3: In $\mathscr{C}$, the temporal constraint $1[l : u]3$ would mean that B must occur within $[l : u]$ of the occurrence of lead A.[5] In $\mathscr{C}'$, the very same temporal constraint $1[l : u]3$ would mean that C must occur within $[l : u]$ of the occurrence of lead A.

The relation $\in$ is a partial order on the set of multisets over a set X (see for instance Lemma 4.23 in [62]). It is of interest to check whether a notion of intersection coincides with the notion of greatest lower bound for $\in$.

Definition 2.7 The intersection of two multisets $\mathcal{E}$ and $\mathcal{E}'$ is denoted $\mathcal{E} \cap \mathcal{E}'$ and is given by the multiplicity function $\mu_{\mathcal{E} \cap \mathcal{E}'}(e) = \min(\mu_{\mathcal{E}}(e), \mu_{\mathcal{E}'}(e))$ for all $e \in \mathbb{E}$.

Lemma 2.4 $\mathcal{E} \cap \mathcal{E}'$ is the greatest lower bound of $\mathcal{E}$ and $\mathcal{E}'$ wrt $\in$.

2.3   A Preorder Between Chronicles

Before defining a preorder on the set of chronicles, we need to define a relation between the temporal constraints.

Notation We write $(e_1, o_1)[l : u](e_2, o_2) \trianglelefteq (e'_1, o'_1)[l' : u'](e'_2, o'_2)$ iff $e_1 = e'_1$, $e_2 = e'_2$ and $[l : u] \subseteq [l' : u']$.

Please observe that $\trianglelefteq$ involves no condition upon positions (i.e., no condition upon o_i and o'_i). This means that the relation holds between temporal constraints regardless to the multiset events except for e_1, e_2 or e'_1, e'_2.

Now, we can define a relation on the set of chronicles.

Definition 2.8 (Preorder on Chronicles) Let $\mathscr{C} = (\mathcal{E}, \mathcal{T})$ and $\mathscr{C}' = (\mathcal{E}', \mathcal{T}')$ be chronicles. $\mathscr{C}$ is **at least as specific as** $\mathscr{C}'$, to be denoted $\mathscr{C}' \preceq \mathscr{C}$, iff $\mathcal{E}' \in \mathcal{E}$ by some multiset embedding θ such that for each $(e'_i, o)[l' : u'](e'_j, p)$ in $\mathcal{T}'$ there is a temporal constraint $(e_{\theta(i)}, \theta(o))[l : u](e_{\theta(j)}, \theta(p))$ in $\mathcal{T}$ satisfying[6] $(e_{\theta(i)}, \theta(o))[l : u](e_{\theta(j)}, \theta(p)) \trianglelefteq (e'_i, o)[l' : u'](e'_j, p)$.

Remark The symbol $\preceq$ is to be read as "is less specific than (or as specific as)". In the special case that $\mathscr{C}' \preceq \mathscr{C}$ but not vice-versa, it can be said, according to Definition 2.8, that $\mathscr{C}'$ is *less specific than* $\mathscr{C}$.

The *less specific* may come from two different reasons: because the multiset is *smaller* and/or because the temporal constraints are *wider*. Intuitively, a chronicle is

---

[5] We use the phrase "occurrence of lead A" to make the difference from the "first occurrence of A" that we will introduce in Chap. 4. In general, the two notions do actually differ.

[6] Since θ is a multiset embedding for $\mathcal{E}'$ into $\mathcal{E}$, both $e'_i = e_{\theta(i)}$ and $e'_j = e_{\theta(j)}$.

Fig. 2.6 Two incomparable chronicles (Example 2.4)

more specific that another if the former specifies more events or if it has *narrowed* temporal constraints.

Lemma 2.5 $\preceq$ *is a preorder.*

Notation We write $\mathscr{C} \approx \mathscr{C}'$ (i.e., $\mathscr{C}$ and $\mathscr{C}'$ are equivalent) iff $\mathscr{C} \preceq \mathscr{C}'$ and $\mathscr{C}' \preceq \mathscr{C}$.

Example 2.4 (Two Incomparable Chronicles) The chronicles (depicted in Fig. 2.6)

$$\mathscr{C} = \left(\{\!\{ABB\}\!\}, \left\{ \begin{matrix} (A, 1)[1:2](B, 2), \\ (A, 1)[5:6](B, 3) \end{matrix} \right\} \right)$$

and

$$\mathscr{C}' = \left(\{\!\{AB\}\!\}, \left\{ \begin{matrix} (A, 1)[1:4](B, 2), \\ (A, 1)[3:6](B, 2) \end{matrix} \right\} \right)$$

are such that $\mathscr{C} \npreceq \mathscr{C}'$ (for the reason that $\{\!\{ABB\}\!\} \not\subseteq \{\!\{AB\}\!\}$) and $\mathscr{C}' \npreceq \mathscr{C}$ (this is less obvious, it comes from the fact that *two* embeddings are needed to compare $(A, 1)[1 : 2](B, 2)$ with $(A, 1)[1 : 4](B, 2)$ on the one hand and $(A, 1)[5 : 6](B, 3)$ with $(A, 1)[3 : 6](B, 2)$ on the other hand. Indeed, Definition 2.8 does not allow the multiset embedding used for $\trianglelefteq$ to vary (even in the case, as here, where there are more than one multiset embedding for $\mathcal{E}' \in \mathcal{E}$). However, if either temporal constraint in $\mathscr{C}'$ is dropped then only one multiset embedding is needed and $\mathscr{C}' \preceq \mathscr{C}$ then holds.

That $\preceq$ is only a pre-order but not a partial order is shown by exhibiting two distinct chronicles $\mathscr{C}$ and $\mathscr{C}'$ such that $\mathscr{C} \approx \mathscr{C}'$ as in Example 2.5 (see Fig. 2.7).

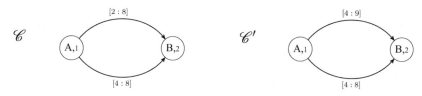

Fig. 2.7 Two equivalent but distinct chronicles: $\mathscr{C} \approx \mathscr{C}'$ and $\mathscr{C} \neq \mathscr{C}'$ (Example 2.5)

Example 2.5 (Two Chronicles $\mathscr{C} \neq \mathscr{C}'$ such that $\mathscr{C} \preceq \mathscr{C}'$ and $\mathscr{C}' \preceq \mathscr{C}$) The chronicles (depicted in Fig. 2.7)

$$\mathscr{C} = \left(\{\{AB\}\}, \left\{ \begin{array}{l} (A, 1)[2:8](B, 2), \\ (A, 1)[4:8](B, 2) \end{array} \right\} \right)$$

and

$$\mathscr{C}' = \left(\{\{AB\}\}, \left\{ \begin{array}{l} (A, 1)[4:9](B, 2), \\ (A, 1)[4:8](B, 2) \end{array} \right\} \right)$$

are such that $\mathscr{C} \neq \mathscr{C}'$ and $\mathscr{C} \approx \mathscr{C}'$.

Note that we have $\mathscr{C}' \preceq \mathscr{C}$, because each temporal constraint of $\mathscr{C}'$ is larger than the ones of $\mathscr{C}$ (according to $\trianglelefteq$). On the opposite, the two temporal constraints of $\mathscr{C}$ are larger than the temporal constraint $(A, 1)[4:8](B, 2)$ of $\mathscr{C}'$. This is sufficient to have $\mathscr{C} \preceq \mathscr{C}'$ even if none of the temporal constraints of $\mathscr{C}$ are larger than $(A, 1)[4:9](B, 2)$ in $\mathscr{C}'$.

A consequence of Example 2.3 is that $\mathcal{E}' \Subset \mathcal{E}$ fails to imply $(\mathcal{E}', \mathcal{T}) \preceq (\mathcal{E}, \mathcal{T})$. Instead, a weaker property holds as follows.

Lemma 2.6 *Let $\mathscr{C} = (\mathcal{E}, \mathcal{T})$ and $\mathscr{C}'' = (\mathcal{E}'', \mathcal{T}'')$ be chronicles such that $\mathscr{C}'' \preceq \mathscr{C}$. If $\mathcal{E}'' \Subset \mathcal{E}' \Subset \mathcal{E}$ then $(\mathcal{E}', \mathcal{T}'') \preceq \mathscr{C}$.*

Definition 2.8 expresses that any two temporal constraints in a chronicle are interpreted conjunctively: $(\mathcal{E}, \mathcal{T}) \preceq (\mathcal{E}, \mathcal{T} \cup \mathcal{T}')$. That is, supplementing a chronicle with extra temporal constraints gives a more specific chronicle.

Lemma 2.7 $(\mathcal{E}, \mathcal{T}') \preceq (\mathcal{E}, \mathcal{T})$ *whenever $\mathcal{T}' \subseteq \mathcal{T}$.* $\qquad\qquad\square$

Lemma 2.8 *If $(\mathcal{E}, \mathcal{T}') \preceq (\mathcal{E}, \mathcal{T})$ then $(\mathcal{E}, \mathcal{T} \cup \mathcal{T}') \preceq (\mathcal{E}, \mathcal{T})$.*

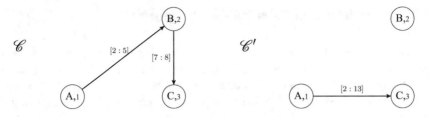

Fig. 2.8 $(\mathcal{E}, \mathcal{T}') \npreceq (\mathcal{E}, \mathcal{T})$ despite $(\mathcal{E}, \mathcal{T}') \preceq (\mathcal{E}, \mathcal{T}^*)$ (Example 2.6)

An instance of Lemma 2.5 worth mentioning is about unlimited temporal constraints (Definition 2.5), that is:

$$(\mathcal{E}, \mathcal{T}) \preceq (\mathcal{E}, \mathcal{T} \cup \{(e, o_e)[-\infty : +\infty](e', o_{e'})\})$$

The converse holds in the case that there exists some $(e, o_e)[l : u](e', o_{e'})$ in $\mathcal{T}$ (with either $l \neq -\infty$ or $u \neq +\infty$), and that is:

$$(\mathcal{E}, \mathcal{T} \cup \{(e, o_e)[-\infty : +\infty](e', o_{e'})\}) \preceq (\mathcal{E}, \mathcal{T})$$

As to (pre-)ordering chronicles, $\preceq$ is rather weak since it compares sets of temporal constraints instead of the ∗-version (see the closure $\mathcal{T}^*$ in Definition 2.4) thereof. Here is an illustration (see also Fig. 2.8).

Example 2.6 (($\mathcal{E}, \mathcal{T}') \preceq (\mathcal{E}, \mathcal{T}^*)$ fails to entail $(\mathcal{E}, \mathcal{T}') \preceq (\mathcal{E}, \mathcal{T}))$ Consider $\mathcal{E} = \{\!\{ABC\}\!\}$. Let

$$\mathcal{T}' = \{(A, 1)[2 : 13](C, 3)\}$$

i.e., $\mathcal{T}'$ demands that $(C, 3)$ occurs in the interval $[2 : 13]$ wrt $(A, 1)$. Let

$$\mathcal{T} = \left\{ \begin{array}{l} (A, 1)[2 : 5](B, 2) \\ (B, 2)[7 : 8](C, 3) \end{array} \right\}$$

hence $\mathcal{T}$ demands that $(C, 3)$ occurs in the interval $[9 : 13]$ wrt $(A, 1)$. But it is not the case that $(\mathcal{E}, \mathcal{T}') \preceq (\mathcal{E}, \mathcal{T})$ (because $(A, 1)[9 : 13](C, 3)$ is not explicitly in $\mathcal{T}$). All this is depicted in Fig. 2.8.

2.4 Chronicles and Subgraph Isomorphisms

Using terminology from the literature on graph mining [45], Definition 2.8 determines a special case of subgraph isomorphism (vertices of domain graph are injectively mapped to vertices of codomain graph such that adjacent vertices receive adjacent images, with the direction of edges being preserved). Fundamentally, it is a special case not because of the labels (indices for vertices and time intervals for edges) which are simply add-ons, but because of the extra requirement in Definition 2.6 where a multiset embedding is required to be strictly increasing instead of being simply required to be injective.

As already indicated, there need not be a unique multiset embedding of $\mathcal{E}'$ into $\mathcal{E}$ (i.e., if there are multiple instances of some event type in $\mathcal{E}$, then there can be more than one way to arrange for a possible counterpart to an event of $\mathcal{E}'$, giving several groundings for the fact that $\mathscr{C}$ is at least as specific as $\mathscr{C}'$).

Reminder In a nutshell, a graph homomorphism is a map between two graphs such that adjacent vertices receive adjacent images. That is, a graph homomorphism preserves edges (arcs in the case of digraphs). See for instance [43] and [30].

Definition Let G_1 and G_2 be digraphs. A graph homomorphism of G_1 to G_2 is a function $h : V_{G_1} \to V_{G_2}$ such that $(x, y) \in E_{G_1} \Rightarrow (h(x), h(y)) \in E_{G_2}$.

If edges and/or vertices have labels, a graph homomorphism preserves not only edges but also all labels. The Example 2.7 illustrates this.

Example 2.7 (Graph Homomorphism)
  Consider the two digraphs below.

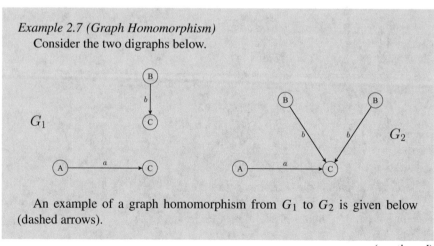

  An example of a graph homomorphism from G_1 to G_2 is given below (dashed arrows).

(continued)

Example 2.7 (continued)

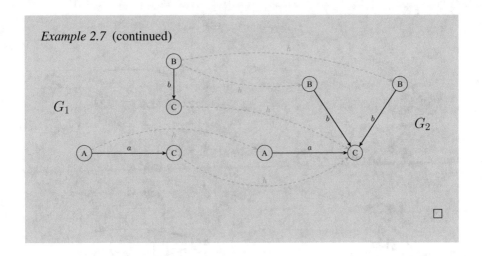

Of special interest are subgraph isomorphisms, i.e., graph homomorphisms that happen to be injective. That is, the existence of such a graph homomorphism shows that the input graph is isomorphic to some subgraph of the target graph. Please observe that the above graph homomorphism from G_1 to G_2 fails to be a subgraph isomorphism: the vertices labelled C in G_1 can only be mapped to the same vertex in G_2, that is, the mapping is not injective.

The view of chronicles as patterns resembling digraphs differs from the issue of subgraph isomorphism when it comes to labels: The condition that labels must be preserved is weakened to containment of temporal intervals. For ℓ the labelling function, $\ell(\langle h(x), h(y) \rangle) \subseteq \ell(\langle x, y \rangle)$ is now required, meaning that the interval labelling the target arc must be contained in the interval labelling the input arc.

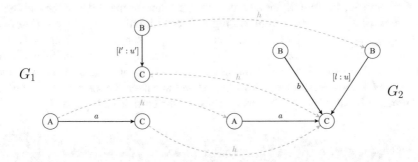

In addition to the previous requirements for h to be a graph homomorphism, it must also be the case that $[l : u] \subseteq [l' : u']$.

However, a chronicle, in general, is not a labelled digraph but a labelled multidigraph, i.e., a digraph allowing for multiple arcs between two vertices and moreover having labelled vertices and arcs. In symbols, a labelled multidigraph is of the form $G = (V, E, L, \ell_V, \ell_E)$ where V is a set of vertices, $E : V \times V \to \mathbb{N}$ gives the number of arcs for each ordered pair of vertices, L is a set of labels,

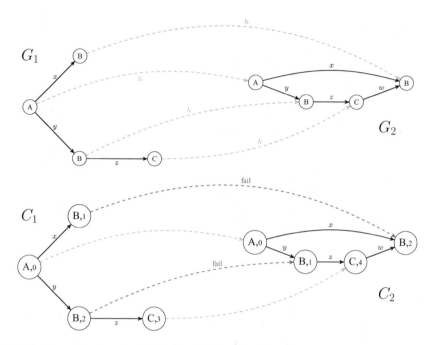

Fig. 2.9 When indexing of nodes blocks embedding (see Example 2.8)

$\ell_V : V \to L$ maps vertices to their label and $\ell_E : V \times V \to \mathbb{N}^L$ such that $\forall a, b \in V$, $E(a, b) = \sum_{l \in L} \ell_E(a, b)(l)$ maps arcs to their label (every two vertices a, b are mapped to a multiset of labels, its cardinality is $E(a, b)$ so that each arc from a to b is assigned exactly one label).

Furthermore, a chronicle has an order over its nodes (as the nodes of the chronicle are indexed) and this makes embedding between chronicles even more different from graph homomorphisms. If there is a subgraph isomorphism from some G_1 to some G_2 then there exists an indexing for the nodes of G_1 and an indexing for the nodes of G_2 such that the resulting chronicle G_1' is embedded in the resulting chronicle G_2'. However, some indexings may fail to extend the relationship between the graphs G_1 and G_2 to a relationship between the chronicles G_1' and G_2'.

Example 2.8 (Graph Homomorphism vs Chronicle Embedding)
   The graph G_1 in Fig. 2.9 does *not* say which of the B-node goes to the C-node whereas the chronicle C_1 imposes that the B-node with lower index goes to the C-node. As a consequence, a subgraph isomorphism exists from G_1 to G_2 but there is no embedding of the chronicle C_1 into the chronicle C_2 because the relative order for the indices of the B-nodes is not preserved (see Fig. 2.9).

Chapter 3
Structuring the Space of Chronicles

Abstract This section is devoted to the definition of subclasses of chronicles with potentially interesting properties for algorithmic usages. More especially, we would like to identify semi-lattice structure in subclasses of chronicles. This section successively introduces the notion of slim, simple and pairwise flush chronicles. The slim chronicle is briefly introduced as a technical construction. Then, we introduce the class of simple chronicles to discard this construction. The class of simple chronicle corresponds to the notion of chronicle commonly encountered in pattern mining literature. We illustrate that we fail to derive a usable construction from this class of chronicles. Then, we propose the notion of flush, which can be used to derive subclasses of chronicles being semi-lattices.

Keywords Simple chronicles · Slim chronicles · Flush chronicles · Poset · Intersection · Semilattice

3.1 Introduction

As the proposed structure on the set of chronicles appears to be too weak regarding our objective to formalize situation recognition and temporal sequence abstraction, this chapter investigates different subsets of chronicles. For these different subsets, we give their definition and propose to structure their space using relation orders and studying their respective properties.

Our journey through the different subsets of chronicles starts with *slim chronicles* which is a technical construct for pursuing the formal account. Then, we introduce *simple chronicles* which denotes the space of chronicle that is usually considered in the literature but for which we illustrate that our attempt for defining a semi-lattice fails. This leads us to investigate new subsets of chronicles and a new manner to structure their space. Section 3.5 introduces a new intersection operator from which a relation order is derived. Under some more restrictions, designing *pairwise flush chronicles*, we finally identify a semi-lattice structure on such subsets of chronicles.

3.2   A Partial Order Between Slim Chronicles

In this section, we introduce the notion of *slim chronicle*. In the previous section, the notion of *well-behaved* chronicle deal with the inconsistent temporal constraints that can be specified in a chronicle. The notion to be introduced now is a starting point to deal with redundancy within a chronicle: discarding chronicles such that a temporal constraint in the chronicle amounts to a condition subsumed by another temporal constraint in the chronicle. Example 2.5 illustrates two distinct but equivalent chronicles for which there are such redundancies. We expect to leverage from suppressing such redundancies for deriving better properties.

Definition 3.1 (Slim Chronicle)   A chronicle $\mathscr{C} = (\mathcal{E}, \mathcal{T})$ is a **slim chronicle** iff for any two distinct temporal constraints $(e_i, o)[t_1^- : t_1^+](e_j, p)$ and $(e_i, o)[t_2^- : t_2^+](e_j, p)$ in $\mathcal{T}$, $[t_1^- : t_1^+] \not\subseteq [t_2^- : t_2^+]$.

A slim chronicle *does* allow for more than one temporal constraint between (e_i, i) and (e_j, j), provided there is no nesting between two of the intervals mentioned in these temporal constraints.

Notation   We write $\widetilde{\mathcal{C}}$ for the set of all slim chronicles over $\mathbb{E}$ and $\mathbb{T}$.

Proposition 3.1 $(\widetilde{\mathcal{C}}, \preceq)$ *is a poset.*

The rationale behind the notion of a slim chronicle can be expressed as follows. Given a chronicle $\mathscr{C} = (\mathcal{E}, \mathcal{T})$, consider a temporal constraint $(e_i, i)[l : u](e_j, j)$ in $\mathcal{T}$. A chronicle $\mathscr{C}' = (\mathcal{E}, \mathcal{T} \cup \{(e_i, i)[l' : u'](e_j, j)\})$ can be defined for $[l : u] \subsetneq [l' : u']$. Then, $\mathscr{C} \approx \mathscr{C}'$.[1] The way $\mathscr{C}'$ is constructed offers grounds to discard one of $\mathscr{C}, \mathscr{C}'$, and that is: Do the inverse of what has been done to get $\mathscr{C}'$, i.e., remove a temporal constraint whose interval is larger than that of another temporal constraint involving exactly the same positions i and j in the multiset.

Since slim chronicles only take care of the most obvious form of redundancy, why stop there? In addition to the computational cost involved in checking whether a chronicle is redundant, a major reason is that slim chronicles form a set of representatives for the equivalence classes induced by $\approx$ as is shown by means of Lemmas 3.4 and 3.5.

Definition 3.2   Let $\mathscr{C} = (\mathcal{E}, \mathcal{T})$ be a chronicle. Define

$$\widetilde{\mathcal{T}} = \mathcal{T} \setminus \{(e_i, o)[l : u](e_j, p) \in \mathcal{T} \mid \exists (e_i, o)[l' : u'](e_j, p) \in \mathcal{T}, \ [l' : u'] \subsetneq [l : u]\}.$$

Also, we write $\widetilde{\mathscr{C}} = (\mathcal{E}, \widetilde{\mathcal{T}})$.

Lemma 3.1 $\widetilde{\mathscr{C}}$ *is a slim chronicle.*   □

---

[1] This can be proved as follows. By Lemma 2.7, $\mathscr{C} \preceq \mathscr{C}'$. As to $\mathscr{C}' \preceq \mathscr{C}$, it results from taking θ to be identity so that the condition in Definition 2.8 holds for every temporal constraint in $\mathcal{T}$ while $(e_i, i)[l : u](e_j, j) \trianglelefteq (e_i, i)[l' : u'](e_j, j)$ is ensured by $[l : u] \subsetneq [l' : u']$.

This lemma indicates that $\widetilde{}$ is an operator to transform a chronicle into a slim chronicle. The following lemmas draw the relationships between a chronicle and its slim form.

Lemma 3.2 *If $\mathscr{C}$ is a slim chronicle then $\mathscr{C} = \widetilde{\mathscr{C}}$.* $\qquad\square$

Lemma 3.3 $\widetilde{\mathscr{C}} \preceq \mathscr{C}$.

Lemma 3.4 *Let $\mathscr{C}$ and $\mathscr{C}'$ be two slim chronicles. If $\mathscr{C} \approx \mathscr{C}'$ then $\mathscr{C} = \mathscr{C}'$.*

Lemma 3.5 *If $\mathscr{C} = (\mathcal{E}, \mathcal{T})$ is a finite chronicle then $\mathscr{C} \approx \widetilde{\mathscr{C}}$.*

As a consequence of Lemmas 3.4 and 3.5, finite slim chronicles form a set of representatives, wrt $\approx$, for finite chronicles. However, Example 3.9 shows that Lemma 3.5 cannot be extended to the case that $\mathcal{T}$ is infinite.

Example 3.9 (A Chronicle That Has No Slim Equivalent) The aim is to show that $\mathscr{C} = (\{\{AAB\}\}, \{(A, 2)[1 : 1 + \frac{1}{n+1}](B, 3) \mid n = 1, 2, \ldots\})$ (see Fig. 3.1) is a chronicle which is equivalent to no slim chronicle. Indeed, assume $\mathscr{C}'$ to be a slim chronicle such that $\mathscr{C} \approx \mathscr{C}'$. In order to have $\mathscr{C}' \preceq \mathscr{C}$, for every $(A, 2)[l : u](B, 3)$ in $\mathscr{C}'$, there must exist $(A, 2)[1 : 1 + \frac{1}{n+1}](B, 3)$ in $\mathscr{C}$ such that $[1 : 1 + \frac{1}{n+1}] \subseteq [l : u]$. Thus, $u > 1$ for every $(A, 2)[l : u](B, 3)$ in $\mathscr{C}'$. On the other hand, $\mathscr{C} \preceq \mathscr{C}'$ does require that, for every $(A, 2)[1 : 1 + \frac{1}{n+1}](B, 3)$, there must exist some $(A, 2)[l : u](B, 3)$ in $\mathscr{C}'$ such that $[l : u] \subseteq [1 : 1 + \frac{1}{n+1}]$. Since $l \leq u$ is ensured by Definition 2.2, it follows that $u \leq 1$ for at least one temporal constraint $(A, 2)[l : u](B, 3)$ in $\mathscr{C}'$. A contradiction arises in view of $u > 1$ for every $(A, 2)[l : u](B, 3)$ in $\mathscr{C}'$.

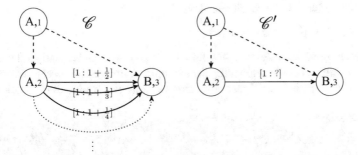

Fig. 3.1 A chronicle $\mathscr{C}$ failing to have a slim equivalent (Example 3.9). There is no upper bound for $\mathscr{C}'$ that would lead to an equivalent chronicle to $\mathscr{C}$

Notation We write $\widetilde{C}_f$ for the set of all finite slim chronicles over $\mathbb{E}$ and $\mathbb{T}$.

At this stage of our formal account, slim chronicles suppress redundancy in temporal constraints and thus enables to have a partially ordered set. The next section investigates a subclass of chronicles we call simple chronicles in order to identify a semi-lattice on some kind of chronicles.

3.3 Simple Chronicles

In this section, we introduce the notion of a simple chronicle in our account. In a nutshell, a simple chronicle with at most one edge between every two nodes. This form of chronicles is traditionally used for chronicle mining algorithms. In these algorithms [2, 16, 21, 22, 27, 63], chronicles do not have multiple edges between the same pair of nodes.

We will see that, surprisingly, this class of chronicles does not have the properties usually required for developing pattern mining algorithms, i.e., a structure of semi-lattice.

Definition 3.3 (Simple Chronicle) A **simple chronicle** $\mathscr{C} = (\mathcal{E}, \mathcal{T})$ is such that for all i, j, l_1, l_2, u_1, u_2, if $(e_i, i)[l_1 : u_1](e_j, j)$ and $(e_i, i)[l_2 : u_2](e_j, j)$ are in $\mathcal{T}$ then they are the same temporal constraint (in symbols, $l_1 = l_2$ and $u_1 = u_2$).

Notation We write $\widehat{C}$ for the set of all simple chronicles over $\mathbb{E}$ and $\mathbb{T}$.

Lemma 3.6 $\widehat{C} \subsetneq \widetilde{C}_f$.   □

Lemma 3.6 expresses that simple chronicles are obviously finite slim chronicles. A counterexample to the converse is, e.g., the chronicle in Fig. 2.1.

We have then a similar result for the space of simple chronicle than for the space of slim chronicle, that is that the space of simple chronicle is a poset with respect to the relation defined in the previous chapter.

Proposition 3.2 $(\widehat{C}, \preceq)$ *is a poset.*

The Example 3.10 illustrates that the set of simple chronicles equipped with $\preceq$ is not a semi-lattice. More specifically, the example exhibits a simple case where there are at least two distinct least general-generalizations of a couple of chronicles.

> *Example 3.10 (Counter-Example)* Let us consider the two chronicles illustrated in Fig. 3.2:
>
> $$\mathscr{C} = (\{\{ABB\}\}, \{(A, 1)[10 : 12](B, 2), (A, 1)[13 : 15](B, 3)\})$$

(continued)

Example 3.10 (continued)
$$\mathscr{C}' = (\{\{AB\}\}, \{(A, 1)[11 : 14](B, 2)\})$$

Then, the two following chronicles are distinct least general-generalizations:

$$lgg_1 = (\{\{AB\}\}, \{(A, 1)[10 : 14](B, 2)\})$$

$$lgg_2 = (\{\{AB\}\}, \{(A, 1)[11 : 15](B, 2)\})$$

It is clear that lgg_1 and lgg_2 are two distinct simple chronicles, and none of them is greater than the other: $lgg_1 \not\preceq lgg_2$ nor $lgg_2 \not\preceq lgg_1$.

But, these two simple chronicles generalize $\mathscr{C}$ and $\mathscr{C}'$. It is simple to see that $lgg_1 \preceq \mathscr{C}'$ (resp. $lgg_2 \preceq \mathscr{C}'$) as there is only one multiset embedding (the identity), and that $[11 : 14] \subseteq [10 : 14]$ (resp. $[11 : 14] \subseteq [11 : 15]$). We also have $lgg_1 \preceq \mathscr{C}$ (resp. $lgg_2 \preceq \mathscr{C}$), because $(A, 1)[10 : 12](B, 3) \trianglelefteq (A, 1)[10 : 14](B, 2)$ (resp. $(A, 1)[13 : 15](B, 2) \trianglelefteq (A, 1)[11 : 15](B, 2)$)

Then, in this case, there is no least general-generalizations.

This example points out the problem of multiple embeddings when dealing with sequential patterns. There are two possible embeddings of $\{\{AB\}\}$ in the multiset $\{\{ABB\}\}$. In this case, each embedding of the multiset $\{\{AB\}\}$ in $\mathscr{C}$ leads to a chronicle that is less specific than $\mathscr{C}$.

Remark Note that this negative result is interesting because simple chronicle is the notion that is usually encountered in the pattern mining literature. Nonetheless, the set of chronicles does not have the usual structure of a pattern set, i.e., it is not a semi-lattice.

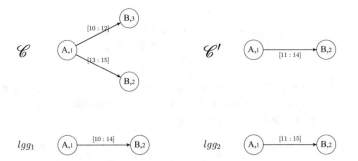

Fig. 3.2 A couple of simple chronicles of Example 3.11 and their two distinct least general-generalizations. It illustrates that $(\widehat{\mathcal{C}}, \preceq)$ is not a semi-lattice

It does not mean that the state of the art algorithms are incorrect, nor incomplete, but it warns that designing an efficient[2] chronicle mining algorithm requires attention.

At this stage of our investigation, our intuition is that the notion of simple chronicle is misleading to identify a semi-lattice structure. Simple chronicles are implicitly associated to a definition of an intersection between two chronicles that consider all the interval of temporal constraints of all possible embeddings to generate a temporal constraint (see details in Sect. B.1 in Appendix).

In the following, we introduce an intersection operator between two chronicles. This intersection will guide us to the definition of subspaces of chronicles for which a semi-lattice structure is possible.

3.4 Intersection Between Chronicles

This section is devoted to a notion of intersection to play the role of greatest lower bound wrt $\preceq$ (when a partial order, i.e., restricted to slim chronicles).

Definition 3.4 (Joint Intersection of Chronicles) Let $\mathscr{C} = (\mathcal{E} = \{\!\{e_1, \ldots, e_m\}\!\},$ $\mathcal{T})$ and $\mathscr{C}' = (\mathcal{E}' = \{\!\{e_1', \ldots, e_{m'}'\}\!\}, \mathcal{T}')$ be chronicles. Let Θ be the set of all multiset embeddings θ for $(\mathcal{E} \cap \mathcal{E}') \Subset \mathcal{E}$ and Θ' the set of all multiset embeddings θ' for $(\mathcal{E} \cap \mathcal{E}') \Subset \mathcal{E}'$. Define the **joint intersection** of $\mathscr{C}$ and $\mathscr{C}'$, denoted $\mathscr{C} \curlywedge \mathscr{C}' = (\mathcal{E} \cap \mathcal{E}', \mathcal{T} \odot \mathcal{T}')$, as

- $\mathcal{E} \cap \mathcal{E}' = \{\!\{e_1'', \ldots, e_s''\}\!\}$ is the multiset intersection of $\mathcal{E}$ and $\mathcal{E}'$
- $\mathcal{T} \odot \mathcal{T}' = \Big\{ (e_i'', i)[\min(l, l') : \max(u, u')](e_j'', j) \mid \mathsf{Proviso} \Big\}$
  where $\mathsf{Proviso}$ stands for

$$\theta \in \Theta,$$
$$\theta' \in \Theta',$$
$$1 \leq i < j \leq s,$$
$$(e_i'', \theta(i))[l : u](e_j'', \theta(j)) \in \mathcal{T},$$
$$(e_i'', \theta'(i))[l' : u'](e_j'', \theta'(j)) \in \mathcal{T}'.$$

Remark The wording "joint intersection" is meant to provide a difference with "split intersection" to be defined in Sect. A.1. In the remainder, "intersection" with no adjective is to be read "joint intersection".

---

[2] i.e., non-redundant, with effective early pruning of the search space.

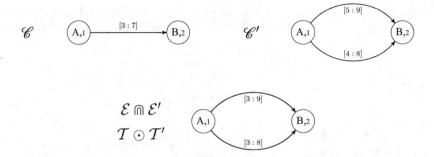

Fig. 3.3 Intersection of two chronicles: the simple case of Example 3.11

Intuitively, the elements of $\mathcal{T} \odot \mathcal{T}'$ in Definition 3.4 can be obtained as follows. First, look at all the pairs (τ, τ') from $\mathcal{T} \times \mathcal{T}'$ such that τ and τ' (as expressed through some $\theta \in \Theta$ and $\theta' \in \Theta'$) happen to impose a temporal condition over the same[3] pair of events:

$$\mathcal{T} \left\{ \begin{array}{c} \vdots \\ (e_{\theta(i)}, \theta(i))[l : u](e_{\theta(j)}, \theta(j)) \\ \vdots \end{array} \right\} \qquad \left\{ \begin{array}{c} \vdots \\ (e_{\theta'(i)}, \theta'(i))[l' : u'](e_{\theta'(j)}, \theta'(j)) \\ \vdots \end{array} \right\} \mathcal{T}'$$

For each such pair, it is the case (see Footnote 3) that $e_{\theta(i)} = e'_{\theta'(i)} = e''_i$ (similarly for j) and $\mathcal{T} \odot \mathcal{T}'$ is then to include the following constraint:

$$(e''_i, i)[\min(l, l') : \max(u, u')](e''_j, j)$$

We now give three examples to support the reader to get deep insights about the proposed intersection operators. A graphical illustration for Definition 3.4 is provided by Fig. 3.3, depicting Example 3.11. Next, Example 3.12 illustrates a more complex joint intersection. Finally, Example 3.13 an interesting particular case where the intersection of a chronicle with looking similar chronicles have very different results.

---

[3] Since θ is a multiset embedding for $\left(\mathcal{E} \cap \mathcal{E}'\right) \in \mathcal{E}$ and θ' a multiset embedding for $\left(\mathcal{E} \cap \mathcal{E}'\right) \in \mathcal{E}'$, Definition 2.6 gives that $e_{\theta(i)} = e''_i$ and $e'_{\theta'(i)} = e''_i$ (it can be the case that $\theta(i) \neq \theta'(i)$, however). Naturally, the same holds for j.

Example 3.11 (Intersection of Two Simple Chronicles) Consider the two chronicles depicted in Fig. 3.3, that is, $\mathscr{C} = (\mathcal{E}, \mathcal{T})$ where

$$\mathcal{E} = \{\!\{AB\}\!\}$$

$$\mathcal{T} = \{\ (A, 1)[3:7](B, 2)\ \}$$

and $\mathscr{C}' = (\mathcal{E}', \mathcal{T}')$ where

$$\mathcal{E}' = \{\!\{AB\}\!\}$$

$$\mathcal{T}' = \{\ (A, 1)[4:8](B, 2),\ (A, 1)[5:9](B, 2)\ \}$$

Obviously, $\mathcal{E} = \mathcal{E}' = \mathcal{E} \sqcap \mathcal{E}'$. Therefore, the set Θ of all multiset embeddings θ for $(\mathcal{E} \sqcap \mathcal{E}') \subseteq \mathcal{E}$ is $\Theta = \{\mathrm{Id}\}$, and, similarly, $\Theta' = \{\mathrm{Id}\}$. That is,

$$\theta : \begin{matrix} 1 \mapsto 1 \\ 2 \mapsto 2 \end{matrix} \qquad\qquad \theta' : \begin{matrix} 1 \mapsto 1 \\ 2 \mapsto 2 \end{matrix}$$

Since $\mathcal{E} \sqcap \mathcal{E}' = \{\!\{AB\}\!\}$, the notation $\mathcal{E} \sqcap \mathcal{E}' = \{\!\{e_1'', e_2''\}\!\}$ used in Definition 3.4 gives $e_1'' = A$ and $e_2'' = B$ (also, $s = 2$). According to Fact 1 on p. 20, θ and θ' can be viewed as the mappings below:[4]

$$\theta : \begin{matrix} e_1'' \mapsto e_1 \\ e_2'' \mapsto e_2 \end{matrix} \qquad\qquad \theta' : \begin{matrix} e_1'' \mapsto e_1' \\ e_2'' \mapsto e_2' \end{matrix}$$

Now, $1 \le i < j \le s = 2$ gives $i = 1$ and $j = 2$ as the unique possibility. As was just mentioned, θ can only be identity and so can θ'. Two cases arise,

$$\mathcal{T}\left\{\begin{matrix} \vdots \\ (A, 1)[3:7](B, 2) \\ \vdots \end{matrix}\right\} \qquad \left\{\begin{matrix} \vdots \\ (A, 1)[4:8](B, 2) \\ \vdots \end{matrix}\right\}\mathcal{T}'$$

and

(continued)

---

[4] To give the full details: The only possible θ maps the A of $\mathcal{E} \sqcap \mathcal{E}' = \{\!\{AB\}\!\}$ to the A of $\mathcal{E} = \{\!\{AB\}\!\}$ and maps the B of $\mathcal{E} \sqcap \mathcal{E}' = \{\!\{AB\}\!\}$ to the B of $\mathcal{E} = \{\!\{AB\}\!\}$. Similarly, the only possible θ' maps the A of $\mathcal{E} \sqcap \mathcal{E}' = \{\!\{AB\}\!\}$ to the A of $\mathcal{E}' = \{\!\{AB\}\!\}$ and maps the B of $\mathcal{E} \sqcap \mathcal{E}' = \{\!\{AB\}\!\}$ to the B of $\mathcal{E}' = \{\!\{AB\}\!\}$.

Example 3.11 (continued)

$$\mathcal{T} \left\{ \begin{array}{c} \vdots \\ (A, 1)[3 : 7](B, 2) \\ \vdots \end{array} \right\} \qquad \left\{ \begin{array}{c} \vdots \\ (A, 1)[5 : 9](B, 2) \\ \vdots \end{array} \right\} \mathcal{T}'$$

that give

$$(A, 1)[\min(3, 4) : \max(7, 8)](B, 2)$$

and

$$(A, 1)[\min(3, 5) : \max(7, 9)](B, 2)$$

Finally,

$$\mathcal{T} \odot \mathcal{T}' = \left\{ \begin{array}{c} (A, 1)[3 : 8](B, 2) \\ (A, 1)[3 : 9](B, 2) \end{array} \right\}$$

Example 3.12 (A More Complex Intersection of Two Chronicles) As depicted in Fig. 3.4, the chronicles in this example are $\mathscr{C} = (\mathcal{E}, \mathcal{T})$ where

$$\mathcal{E} = \{\!\{AAABB\}\!\}$$

$$\mathcal{T} = \{ \ (A, 1)[5 : 10](B, 4),$$

$$(A, 2)[3 : 7](B, 5),$$

$$(A, 3)[2 : 9](B, 5) \ \}$$

and $\mathscr{C}' = (\mathcal{E}', \mathcal{T}')$ where

$$\mathcal{E}' = \{\!\{AABB\}\!\}$$

$$\mathcal{T}' = \{ \ (A, 1)[4 : 8](B, 3),$$

$$(A, 2)[1 : 6](B, 4) \ \}$$

To start with, $\mathcal{E} \cap \mathcal{E}' = \{\!\{AABB\}\!\} = \mathcal{E}'$.

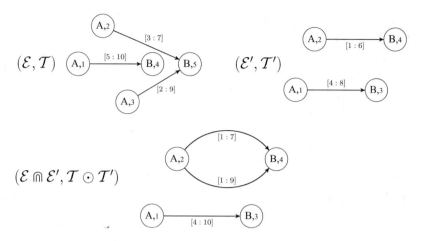

Fig. 3.4 Intersection of two chronicles: a more complex case (see Example 3.12)

Example 3.12 In order to improve readability, a decorated version of Fig. 3.4 is given below (where A is indexed after its position in the multiset of the chronicle and is primed or double-primed depending on whether $\mathcal{E}'$ or $\mathcal{E}''$ is considered).

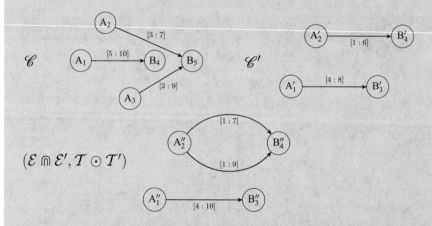

Since $\mathcal{E} \sqcap \mathcal{E}' = \mathcal{E}'$, identity is the only multiset embedding for $\mathcal{E} \sqcap \mathcal{E}' \in \mathcal{E}'$. Thus, the only θ' in Θ' is $\theta' = \text{Id}$.

There are three multiset embeddings for $\mathcal{E} \sqcap \mathcal{E}' \in \mathcal{E}'$. They are

$$
\theta_1 : \begin{matrix} 1 \mapsto 1 \\ 2 \mapsto 2 \\ 3 \mapsto 4 \\ 4 \mapsto 5 \end{matrix} \quad \text{i.e.} \quad \begin{matrix} e_1'' \mapsto e_1 \\ e_2'' \mapsto e_2 \\ e_3'' \mapsto e_4 \\ e_4'' \mapsto e_5 \end{matrix} \qquad \theta_2 : \begin{matrix} 1 \mapsto 1 \\ 2 \mapsto 3 \\ 3 \mapsto 4 \\ 4 \mapsto 5 \end{matrix} \quad \text{i.e.} \quad \begin{matrix} e_1'' \mapsto e_1 \\ e_2'' \mapsto e_3 \\ e_3'' \mapsto e_4 \\ e_4'' \mapsto e_5 \end{matrix}
$$

(continued)

Example 3.12 (continued)

$$\theta_3 : \begin{matrix} 1 \mapsto 2 \\ 2 \mapsto 3 \\ 3 \mapsto 4 \\ 4 \mapsto 5 \end{matrix} \quad \text{i.e.} \quad \begin{matrix} e_1'' \mapsto e_2 \\ e_2'' \mapsto e_3 \\ e_3'' \mapsto e_4 \\ e_4'' \mapsto e_5 \end{matrix}$$

Or, using the notation from the decorated version,

$$\theta_1 : \begin{matrix} A_1'' \mapsto A_1 \\ A_2'' \mapsto A_2 \\ B_3'' \mapsto B_4 \\ B_4'' \mapsto B_5 \end{matrix} \qquad \theta_2 : \begin{matrix} A_1'' \mapsto A_1 \\ A_2'' \mapsto A_3 \\ B_3'' \mapsto B_4 \\ B_4'' \mapsto e_5' \end{matrix} \qquad \theta_3 : \begin{matrix} A_1'' \mapsto A_2 \\ A_2'' \mapsto A_3 \\ B_3'' \mapsto B_4 \\ B_4'' \mapsto B_5 \end{matrix}$$

By comparison, and to be exhaustive,

$$\theta' : \begin{matrix} A_1'' \mapsto A_1' \\ A_2'' \mapsto A_2' \\ B_3'' \mapsto B_3' \\ B_4'' \mapsto B_4' \end{matrix}$$

The process consists of searching $\mathcal{T}$ for $(e_{\theta_k(i)}, \theta(i))[l, u](e_{\theta_k(j)}, \theta(j))$ and searching $\mathcal{T}'$ for $(e_{\theta'(i)}', \theta'(i))[l', u'](e_{\theta'(j)}', \theta'(j))$, ranging over $1 \leq i < j \leq 4$.[5]

- $i = 1$ and $j = 2$.
  This is void because $\mathcal{T}'$ has no temporal constraint of the form $(X, 1)[l' : u'](Y, 2)$ (remember, $\theta' = \text{Id}$ hence $\theta'(j) = j$).
- $i = 1$ and $j = 3$.
  θ_1 offers a solution because $\theta_1(i) = 1$ and $\theta_1(j) = 4$ while $\theta'(i) = 1$ and $\theta'(j) = 3$:

$$\mathcal{T} \left\{ \begin{matrix} \vdots \\ (A, 1)[5 : 10](B, 4) \\ \vdots \end{matrix} \right\} \qquad \left\{ \begin{matrix} \vdots \\ (A, 1)[4 : 8](B, 3) \\ \vdots \end{matrix} \right\} \mathcal{T}'$$

θ_2 offers the same solution ($\theta_2(i) = \theta_1(i)$ for $i = 1$ and $\theta_2(j) = \theta_1(j)$ for $j = 3$) whereas θ_3 offers none for these values of i and j because $\theta_3(i) = 2$ and $\theta_3(j) = 4$ but there is no constraint in $\mathcal{T}$ between A_2 and B_4.

(continued)

---

[5] Please keep in mind that $e_{\theta(i)} = e_i'' = e_{\theta'(i)}'$ (same for j) automatically holds by Definition 3.4. Thus, each step in the process computes a temporal constraint between e_i'' and e_j''.

Example 3.12 (continued)

- $i = 1$ and $j = 4$.
  This is void because there is no temporal constraint in $\mathcal{T}'$ between A'_1 and B'_4 (remember, $\theta' = \mathrm{Id}$ hence $\theta'(i) = i$ and $\theta'(j) = j$).
- $i = 2$ and $j = 3$.
  Here, $\theta_1(i) = 2$ and $\theta_1(j) = 4$ but $\mathcal{T}$ contains no temporal constraint of the form $(X, 2)[l' : u'](Y, 4)$. Also, $\theta_2(i) = 3$ and $\theta_2(j) = 4$ but $\mathcal{T}$ contains no temporal constraint of the form $(X, 3)[l' : u'](Y, 4)$. Lastly, $\theta_3(i) = 3$ and $\theta_3(j) = 4$ but $\mathcal{T}$ contains no temporal constraint of the form $(X, 3)[l' : u'](Y, 4)$.
- $i = 2$ and $j = 4$.
  θ_1 offers a solution because $\theta_1(i) = 2$ and $\theta_1(j) = 5$ while $\theta'(i) = 2$ and $\theta'(j) = 4$ and θ_2 also offers a solution due to $\theta_2(i) = 3$ and $\theta_2(j) = 5$:

$$
\mathcal{T}\left\{ \begin{array}{l} \vdots \\ (A, 2)[3 : 7](B, 5) \\ (A, 3)[2 : 9](B, 5) \\ \vdots \end{array} \right\} \qquad \left\{ \begin{array}{l} \vdots \\ (A, 2)[1 : 6](B, 4) \\ \vdots \end{array} \right\} \mathcal{T}'
$$

θ_3 offers the same solution as θ_2 (due to $\theta_3(i) = \theta_2(i)$ and $\theta_3(j) = \theta_2(j)$).
- $i = 3$ and $j = 4$.
  This is void because $\mathcal{T}$ has no temporal constraint of the form $(X, 4)[l' : u'](Y, h)$ (indeed, $\theta_k(3) = 4$ for $k = 1, \ldots, 3$).

Summing up,

$$\iota(1, 3) = \{([5 : 10], [4 : 8])\}$$

$$\iota(2, 4) = \left\{ \begin{array}{l} ([3 : 7], [1 : 6]) \\ ([2 : 9], [1 : 6]) \end{array} \right\}$$

hence

$$\mathcal{T} \odot \mathcal{T}' = \left\{ \begin{array}{l} (A, 1)[4 : 10](B, 3) \\ (A, 2)[1 : 7](B, 4) \\ (A, 2)[1 : 9](B, 4) \end{array} \right\}$$

Please observe this: for $i = 1$ and $j = 4$, A_2 is the higher indexed A-node among $\{A_1, A_2\}$ hence A_2 can thus play the role of A''_2 whereas for $i = 2$ and $j = 4$, A_2 is the lesser indexed A-node among $\{A_2, A_3\}$ hence A_2 can thus play the role of A''_1.

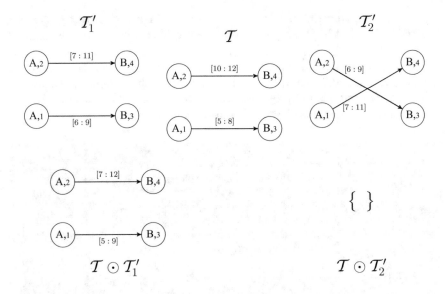

Fig. 3.5 Intersections of closely related chronicles (Example 3.13)

Example 3.13 (Comparison Between the Intersections of Two Chronicles Pairs)
In this example, we consider three chronicles illustrated in Fig. 3.5. Let $\mathcal{C} = (\mathcal{E}, \mathcal{T})$ where

$$\mathcal{E} = \{\!\{AABB\}\!\}$$

$$\mathcal{T} = \{ \ (A, 1)[5 : 8](B, 3), \ \ (A, 2)[10 : 12](B, 4) \ \}$$

Consider $\mathcal{C}_1' = (\mathcal{E}, \mathcal{T}_1')$ where

$$\mathcal{T}_1' = \{ \ (A, 1)[6 : 9](B, 3), \ \ (A, 2)[7 : 11](B, 4) \ \}$$

Consider $\mathcal{C}_2' = (\mathcal{E}, \mathcal{T}_2')$ where

$$\mathcal{T}_2' = \{ \ (A, 1)[7 : 11](B, 4), \ \ (A, 2)[6 : 9](B, 3) \ \}$$

Then, $\mathcal{T} \odot \mathcal{T}_1' = \{(A, 1)[5 : 9](B, 3), \ \ (A, 2)[7 : 12](B, 4)\}$ but $\mathcal{T} \odot \mathcal{T}_2'$ is empty.
  Why is that? In fact, no θ and θ' (multiset embeddings) can map $(A, 1)$ of $\mathcal{C}_1'$ and $(A, 2)$ of $\mathcal{C}_2'$ to the same A-node of $\mathcal{E}$. The reason is that multiset embeddings must preserve the order in the multiset of a chronicle (see again Fig. 2.4). Similarly, $(A, 3)$ of $\mathcal{C}_1'$ and $(A, 1)$ of $\mathcal{C}_2'$ cannot be mapped to the same A-node of $\mathcal{E}$. Of course, this observation also applies with B.

The class of simple chronicles is not closed under the intersection operation (Definition 3.4) that has been proposed in the previous section. An illustration of that is as follows. $\mathscr{C}_1 = (\{\{ABB\}\}, \{(A, 1)[3, 6](B, 2), (A, 1)[7, 9](B, 3)\})$ and $\mathscr{C}_2 = (\{\{AB\}\}, \{(A, 1)[4, 8](B, 2)\})$ are two simple chronicles. By definition, $\mathscr{C}_1 \curlywedge \mathscr{C}_2 = (\{\{AB\}\}, \{(A, 1)[3, 8](B, 2), (A, 1)[4, 9](B, 2)\})$ which is not a simple chronicle.

We complement this result by illustrating (see Example 3.14) that the space of slim chronicles is not event closed under the joint intersection.

Example 3.14 (Intersection of Two Slim Chronicles Is Not Slim)
In this example, we consider two slim chronicles $\mathscr{C} = (\{\{AB\}\}, \mathcal{T})$ and $\mathscr{C}' = (\{\{AB\}\}, \mathcal{T}')$ where:

$$\mathcal{T} = \{ \ (A, 1)[1 : 4](B, 2), \ (A, 1)[2 : 5](B, 2) \ \}$$

$$\mathcal{T}' = \{ \ (A, 1)[3 : 6](B, 2) \ \}$$

Then, $\mathcal{T} \odot \mathcal{T}' = \{ \ (A, 1)[1 : 6](B, 2), \ (A, 1)[2 : 6](B, 2) \ \}$ and $\mathscr{C} \curlywedge \mathscr{C}' = (\{\{AB\}\}, \mathcal{T} \odot \mathcal{T}')$ is not slim.

We are now in position to introduce a sufficient condition for joint intersection to play the role of greatest lower bound wrt $\preceq$ or equivalently, that a lower semi-lattice can be defined upon the set of chronicles. But for that, we need to identify a set of chronicle that will be closed under the joint intersection. With this purpose, the section introduces a new class of chronicles we call *pairwise flush chronicles*.

3.5 Semilattice Structure for Subspaces of the Chronicles

Definition 3.5 Two chronicles $\mathscr{C} = (\mathcal{E}, \mathcal{T})$ and $\mathscr{C}' = (\mathcal{E}', \mathcal{T}')$ are **flush** if they satisfy the condition

$$\forall e \in \mathbb{E} \ \text{if} \ \mu_{\mathcal{E}}(e) \neq 0 \ \text{and} \ \mu_{\mathcal{E}'}(e) \neq 0 \ \text{then} \ \mu_{\mathcal{E}}(e) = \mu_{\mathcal{E}'}(e).$$

Intuitively, two different chronicles are flush when they have different sets of types of events, but the exact same numbers of events on the shared types. Example 3.15 illustrates some flush chronicles.

Example 3.15 (Flush Chronicles) Let $\mathscr{C} = (\mathcal{E} = \{\!\{ABBD\}\!\}, \mathcal{T})$, $\mathscr{C}' = (\mathcal{E}' = \{\!\{ABBC\}\!\}, \mathcal{T}')$, $\mathscr{C}'' = (\{\!\{ABB\}\!\}, \mathcal{T}'')$, $\mathscr{C}''' = (\mathcal{E}''' = \{\!\{BBCDD\}\!\}, \mathcal{T}''')$ be four chronicles.

The following pairs are made of flush chronicles: $(\mathscr{C}'', \mathscr{C})$, $(\mathscr{C}', \mathscr{C})$, $(\mathscr{C}', \mathscr{C})$, $(\mathscr{C}''', \mathscr{C}'')$ and $(\mathscr{C}''', \mathscr{C}')$, but $\mathscr{C}'''$ and $\mathscr{C}$ are not pairwise flush as they both hold D event types but not with the same multiplicity.

Considering the pair $(\mathscr{C}'', \mathscr{C}')$, the multiset of $\mathscr{C}''$ is included into the multiset of $\mathscr{C}'$. The unique embedding is $\{1 \mapsto 1, 2 \mapsto 2, 3 \mapsto 3\}$.

Considering the pair $(\mathscr{C}''', \mathscr{C})$ that is not flush, we have $\mathcal{E}''' \cap \mathcal{E} = \{\!\{BBD\}\!\}$. There are two possible embeddings from is $\mathcal{E}''' \cap \mathcal{E}$ to $\mathcal{E}'''$ (one per D in $\mathcal{E}'''$): $\{1 \mapsto 1, 2 \mapsto 2, 3 \mapsto 4\}$ and $\{1 \mapsto 1, 2 \mapsto 2, 3 \mapsto 5\}$.

We now give some intermediary results for pairs of flush chronicles that will lead us our main result on the structure of the space of pairwise flush chronicles.

Lemma 3.7 *If $\mathscr{C} = (\mathcal{E}, \mathcal{T})$ and $\mathscr{C}' = (\mathcal{E}', \mathcal{T}')$ are flush then $\mathscr{C}$ and $(\mathcal{E} \cap \mathcal{E}', \mathcal{T}'')$ are flush.*

Lemma 3.8 *Let $\mathscr{C} = (\mathcal{E}, \mathcal{T})$ and $\mathscr{C}' = (\mathcal{E}', \mathcal{T}')$ be such that $\mathcal{E}' \Subset \mathcal{E}$. If $\mathscr{C}$ and $\mathscr{C}'$ are flush then there is a single multiset embedding for $\mathcal{E}' \Subset \mathcal{E}$.*

Lemma 3.9 *If $\mathscr{C} = (\mathcal{E}, \mathcal{T})$ and $\mathscr{C}' = (\mathcal{E}', \mathcal{T}')$ are flush then there is a unique multiset embedding for $\mathcal{E} \cap \mathcal{E}' \Subset \mathcal{E}$.*

Lemma 3.10 *If $\mathscr{C}$ and $\mathscr{C}'$ are flush then $\mathscr{C} \curlywedge \mathscr{C}' \preceq \mathscr{C}$.*

Lemma 3.11 *Let $\mathscr{C}, \mathscr{C}', \mathscr{C}''$ be pairwise flush. If $\mathscr{C}''$ is a lower bound of $\mathscr{C}$ and $\mathscr{C}'$ then $\mathscr{C}'' \preceq \mathscr{C} \curlywedge \mathscr{C}'$.*

Remark The reader can notice that the notion of flush is independent from being simple, slim or finite.

There remains to introduce a convenient notion to specify maximal classes of flush chronicles: A profile P is used to set the non-null multiplicity of instances of each event type in a family of chronicles.

Definition 3.6 We call **profile** a map from the event types to the natural numbers. A profile P is **bounded** iff the preimage of 0 is cofinite, i.e., $|\{e \in \mathbb{E} \mid \mu(e) > 0\}| < \infty$.

Notation By definition, a bounded profile P defines a unique multiset. Then the notation P of a bounded profile represents either a map or a multiset.

Definition 3.7 A chronicle $\mathscr{C} = (\mathcal{E}, \mathcal{T})$ **conforms with** a profile P iff for every $e \in \mathbb{E}$, $\mu_{\mathcal{E}}(e) \leq P(e)$. It **conforms exactly with** P iff $\mu_{\mathcal{E}}(e) = P(e)$ for all $e \in \mathbb{E}$.

Notation For a profile P, let $\widetilde{\mathcal{C}_{fP}}$ denote the set of all finite slim chronicles that conform exactly with a profile P.

Proposition 3.3 *For P a bounded profile, $(\widetilde{\mathcal{C}_{fP}}, \preceq)$ is a lower semi-lattice.*

The Proposition 3.3 indicates that there exists families of chronicles that are lower semi-lattices. This families may seem restrictive as it requires defining a profile to which all the chronicles have to conform exactly. Intuitively, it means that all the chronicles of the family have the same finite multiset of event types. Nonetheless, we will see in Chap. 5, that this family enables to have some interesting practical results.

3.6 Summary

In this chapter, we made a journey in some subclasses of chronicles in order to identify a lower semi-lattice. The Fig. 3.6 illustrates the subclasses we investigate and shows their relationships.

The main classes that have been introduced are the following. For each of them, to briefly remind their main characteristics.

- $\widetilde{\mathcal{C}}$ (slim chronicles): they avoid some redundancies between the temporal constraints but forbidding to have inclusion of a temporal constraint in another.
- $\mathcal{C}_f$ (Finite chronicles): they avoid having infinite set of temporal constraints.
- $\widetilde{\mathcal{C}_f}$ (Finite slim chronicles) they combine finiteness and non redundancies of temporal constraints.
- $\widehat{\mathcal{C}}$ (Simple chronicles): they hold at most one temporal constraint between each pair of events.
- $\widetilde{\mathcal{C}_{fP}}$ (Finite slim chronicles pairwise flush): families of finite slim chronicles defined by a profile P that sets the multiset of event types.

All along this chapter, we propose a relation order between chronicles under which the simple chronicles and the slim chronicles are posets.

To go further toward the definition of lower semi-lattices, we investigated two different types of intersections between chronicles. The *simple joint intersection* has

Fig. 3.6 Illustration of the different notions introduced in the chapter. The arrows show the subset relation between sets. The dashed lines figure out that there are multiple families of pairwise flush chronicles and chronicles conformed to profiles P

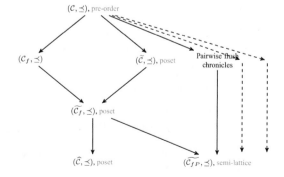

first been derived from the literature in pattern mining. But it turns out to not induce a semi-lattice on the space of simple chronicles (Example 3.10 provides a simple counter-example). Then, we introduced a joint intersection ($\curlywedge$) which enables us to conclude to semi-lattices for the families of pairwise flush chronicles.

Chapter 4
Occurrences of a Chronicle

Abstract This chapter gives elements about how a chronicle occurs in a sequence of events. The seminal work defining the notion of chronicles was motivated by system monitoring applications. A chronicle defines a temporal arrangement of events whose occurrences have to be monitored (for instance, for firing alarms). This section exploits our formal account of chronicles to formalize the notion of occurrence of a chronicle in a sequence. In addition, we introduce the multiple occurrences of a chronicle in a sequence and how to count them.

Keywords Temporal sequence · Embedding · Enumeration

4.1 Introduction

As we have seen in the introduction of this book, a chronicle defines a situation by a temporal arrangement of events—or a pattern. The user is interested in detecting whether or when such a situation happens during the functioning of a system to monitor. For that, it is assumed that the system to monitor yields event logs. Then, the problem of situation recognition with chronicles is to look for occurrences of a chronicle in the event logs. Situation recognition is widely used in monitoring systems, for example, in the context of intrusion detection system or in smart industry. Thus, this task has been widely studied, especially in the field of model checking [11]. A specificity (and the interest) of chronicles is to specify metric temporal constraints on the arrangement of events.

This chapter is focused on the definition of a formal notion of an occurrence of a chronicle in a sequence. And to provide some results deduced from the structure on the set of chronicles.

Let us start by the basic definition of sequences and stripped sequences.

Definition 4.1 A **sequence** (or event sequence) is a pair

$$\langle I, \langle (e_1, t_1), (e_2, t_2), \ldots, (e_n, t_n) \rangle \rangle$$

where I is an identifier distinctive of the sequence and $\langle (e_1, t_1), (e_2, t_2), \ldots, (e_n, t_n) \rangle$ is a finite list of events called **stripped sequence**. The events in the stripped sequence are ordered by $\lessdot$ defined as

$$\forall i, j \in \{1, \ldots, n\}, \; i < j \Leftrightarrow (e_i, t_i) \lessdot (e_j, t_j) \Leftrightarrow t_i < t_j \vee (t_i = t_j \wedge e_i <_{\mathbb{E}} e_j).$$

Lemma 4.1 *In a stripped sequence, no event is repeated.* ☐

Note that this chapter only deals with stripped sequences. The labels will be used in Chap. 5 when we will consider collection of sequences.

4.2 Occurrence of a Chronicle

This section deals with a formal definition underlying the notion of chronicles as patterns to be found in event sequences. The notion of occurrence is a core concept for situation recognition.

Definition 4.2 (Chronicle Occurrences and Embeddings) An **occurrence** of a chronicle $\mathscr{C} = (\{\!\{e'_1, \ldots, e'_m\}\!\}, \mathcal{T})$ in a stripped sequence $S = \langle (e_1, t_1), \ldots, (e_n, t_n) \rangle$ is a subset[1] $\{(e_{f(1)}, t_{f(1)}), \ldots, (e_{f(m)}, t_{f(m)})\}$ of S such that

1. $f : [m] \rightarrow [n]$ is an injective function,
2. $f(i) < f(i+1)$ whenever $e'_i = e'_{i+1}$,
3. $e'_i = e_{f(i)}$ for $i = 1, \ldots, m$,
4. $t_{f(j)} - t_{f(i)} \in [t^- : t^+]$ whenever $(e'_i, i)[t^- : t^+](e'_j, j) \in \mathcal{T}$.

We call f an **embedding**.

The chronicle $\mathscr{C}$ **occurs** in S, denoted $\mathscr{C} \Subset S$, iff there is at least one occurrence of $\mathscr{C}$ in S.

When ordered by $\lessdot$, the subset $\{(e_{f(1)}, t_{f(1)}), \ldots, (e_{f(m)}, t_{f(m)})\}$ happens to be a subsequence of S.

Convention $(\emptyset, \emptyset)$ is considered to occur in all stripped sequences.

It is worth noting that f usually fails to be increasing. In fact, there is a difference between (1) the order $\lessdot$ on the events in the sequence, and (2) the order $\leq_{\mathbb{E}}$ on the event types in the multiset of the chronicle. This is mainly due to the possibility to have negative values in the temporal constraints (see Example 4.2). In fact, (3) conforms to the next lemma.

Lemma 4.2 *If $i < j$ then $e_{f(i)} \leq_{\mathbb{E}} e_{f(j)}$.*

---

[1] Since Lemma 4.1 states that no event is repeated in a stripped sequence, the members of S form a set of events (whose cardinality is the length of S).

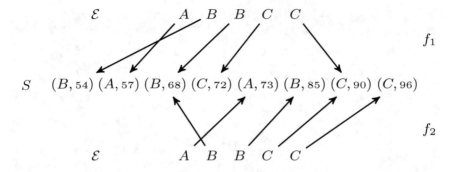

Fig. 4.1 Chronicle occurrences in a stripped sequence (Example 4.16)

As is naturally expected, it can happen that a chronicle has several occurrences in the same stripped sequence: see the next example (depicted in Fig. 4.1).

Example 4.16 (Two Occurrences of a Chronicle in a Stripped Sequence) Let S be the stripped sequence

$$\Big\langle (B, 54), \quad (A, 57), \quad (B, 68), \quad (C, 72), \quad (A, 73), \quad (B, 85), \quad (C, 90), \quad (C, 96)\Big\rangle$$

Consider the chronicle $\mathscr{C} = (\mathcal{E}, \mathcal{T})$ where $\mathcal{E} = \{\!\{ABBCC\}\!\}$ and

$$\mathcal{T} = \left\{ \begin{array}{l} (A, 1)[-6, -1](B, 2) \\ (A, 1)[10, 12](B, 3) \\ (B, 2)[25, 50](C, 4) \\ (B, 3)[3, 7](C, 5) \end{array} \right\}$$

In view of f_1 and f_2,

$$
\begin{array}{ccccc}
 & [m] \to [n] & & \mathcal{E} & \to & S \\
 & 1 \mapsto 2 & & (A, 1) & \rightsquigarrow & (A, 103) \\
f_1: & 2 \mapsto 1 & \text{i.e.} & (B, 2) & \rightsquigarrow & (B, 100) \\
 & 3 \mapsto 3 & & (B, 3) & \rightsquigarrow & (B, 114) \\
 & 4 \mapsto 4 & & (C, 4) & \rightsquigarrow & (C, 118) \\
 & 5 \mapsto 7 & & (C, 5) & \rightsquigarrow & (C, 136)
\end{array}
$$

(continued)

Example 4.16 (continued)

$$[m] \rightarrow [n] \qquad \mathcal{E} \quad \rightarrow \quad S$$

$$
f_2:
\begin{array}{rcl}
1 & \mapsto & 5 \\
2 & \mapsto & 3 \\
3 & \mapsto & 6 \\
4 & \mapsto & 7 \\
5 & \mapsto & 8
\end{array}
\quad \text{i.e.} \quad
\begin{array}{rcl}
(A, 1) & \rightsquigarrow & (A, 119) \\
(B, 2) & \rightsquigarrow & (B, 114) \\
(B, 3) & \rightsquigarrow & (B, 131) \\
(C, 4) & \rightsquigarrow & (C, 136) \\
(C, 5) & \rightsquigarrow & (C, 142)
\end{array}
$$

it is clear that there are two occurrences of $\mathscr{C}$ in S.

Where arrows cross in Fig. 4.1, f_k is not increasing (indeed, Condition 1. in Definition 4.2 precludes crossing of arrows just in case the event type is the same). This is due to the negative bounds of the temporal constraint between $(A, 1)$ and $(B, 2)$.

Proposition 4.1 *If a chronicle $\mathscr{C}$ occurs in at least one stripped sequence, then $\mathscr{C}$ is well-behaved.*

The converse of Proposition 4.1 is untrue as is illustrated by Example 4.17.

Example 4.17 (A Well-Behaved Chronicle that Occurs in No Stripped Sequence) Take

$$\mathcal{E} = \{\!\{ABB\}\!\}$$

and

$$
\mathcal{T} = \left\{
\begin{array}{l}
(A, 1)[5 : 8](B, 2), \\
(A, 1)[2 : 3](B, 3)
\end{array}
\right\}.
$$

We now show that, if $S = \langle (e_1, t_1), \ldots, (e_n, t_n) \rangle$ is a stripped sequence in which $\mathscr{C} = (\mathcal{E}, \mathcal{T})$ occurs, then a contradiction arises. Since $\mathscr{C}$ occurs in S, Definition 4.2 entails that there exists a subset $\{e_{f(1)}, t_{f(1)}), \ldots, (e_{f(m)}, t_{f(m)})\}$ of S such that, for some $f : [m] \rightarrow [n]$ where m is the cardinality of $\mathcal{E}$, all of the following properties hold: (1) f is injective, (2) $f(i) < f(i + 1)$ whenever $e'_i = e'_{i+1}$, (3) $e'_i = e_{f(i)}$ for $i = 1, \ldots, m$ (4) $t_{f(j)} - t_{f(i)} \in [t^- : t^+]$ whenever $(e'_i, i)[t^- : t^+](e'_j, j) \in \mathcal{T}$.

According to Definition 4.2, $m = 3$ here because $\mathcal{E} = \{\!\{e'_1, \ldots, e'_m\}\!\} = \{\!\{ABB\}\!\}$. Also, $e'_1 = A$ and $e'_2 = e'_3 = B$. By (4), $(A, 1)[5 : 8](B, 2) \in \mathcal{T}$ and

(continued)

Proposition 4.2 *Let $\mathcal{C}$ and $\mathcal{C}'$ be two chronicles such that $\mathcal{C}' \preceq \mathcal{C}$. For all stripped sequence S, if $\mathcal{C}$ occurs in S then $\mathcal{C}'$ occurs in S.*

An immediate follow-up of Proposition 4.2, namely Corollary 4.1, provides further justification to slim chronicles. Indeed, a consequence of Corollary 4.1 is that if a slim chronicle $\mathcal{C}'$ is obtained by removing some redundant temporal constraint(s) from a chronicle $\mathcal{C}$ then $\mathcal{C}$ and $\mathcal{C}'$ occur in exactly the same stripped sequences.

Corollary 4.1 *Let $\mathcal{C}$ and $\mathcal{C}'$ be two chronicles such that $\mathcal{C} \approx \mathcal{C}'$. For all stripped sequence S, $\mathcal{C}$ occurs in S iff $\mathcal{C}'$ occurs in S.* $\square$

Converse of Corollary 4.1 is untrue in the sense that two chronicles can occur in exactly the same stripped sequences although they are not equivalent. This is illustrated by Example 4.18 (depicted in Fig. 4.2).

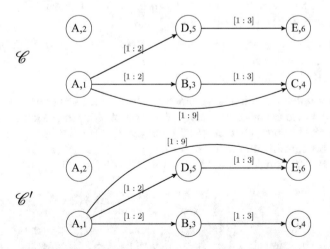

Fig. 4.2 $\mathcal{C} \not\approx \mathcal{C}'$ but $\mathcal{C}$ and $\mathcal{C}'$ occur in the same stripped sequences (Example 4.18)

Example 4.18 (Non-Equivalent Chronicles that Occur in the Same Sequences) Here, we exhibit two slim chronicles $\mathscr{C} = (\mathscr{E}, \mathcal{T})$ and $\mathscr{C}' = (\mathscr{E}', \mathcal{T}')$ (depicted in Fig. 4.2) that occur in exactly the same stripped sequences despite $\mathscr{C} \not\approx \mathscr{C}'$. Consider $\mathscr{E} = \mathscr{E}' = \{\{AABCDE\}\}$. Then, consider $\mathcal{T} = \mathcal{T}'' \cup \{(A, 1)[1 : 9](C, 4)\}$ and $\mathcal{T} = \mathcal{T}'' \cup \{(A, 1)[1 : 9](E, 6)\}$ where

$$
\mathcal{T}'' = \left\{ \begin{array}{l} (A, 1)[1 : 2](B, 3), \\ (B, 3)[1 : 3](C, 4), \\ (A, 1)[1 : 2](D, 5), \\ (D, 5)[1 : 3](E, 6) \end{array} \right\}.
$$

Clearly, $(A, 1)[1 : 9](C, 4)$ plays no role regarding what stripped sequences $\mathscr{C}$ occurs in because the delay between $(A, 1)$ and $(C, 4)$ is actually imposed by the first two temporal constraints of $\mathcal{T}''$ (through $[1 : 2]$ and $[1 : 3]$, they require a tighter interval for the overall delay between $(A, 1)$ and $(C, 4)$). Similarly, $(A, 1)[1 : 9](E, 6)$ plays no role regarding what stripped sequences $\mathscr{C}'$ occurs in because the delay between $(A, 1)$ and $(E, 6)$ is actually imposed by the third and fourth temporal constraints of $\mathcal{T}''$. Summing up, only the temporal constraints in $\mathcal{T}''$ play a role and this means that $\mathscr{C}$ and $\mathscr{C}'$ occur exactly in the same stripped sequences.

Consider the property that for all stripped sequence S, if $\mathscr{C}$ occurs in S then $\mathscr{C}'$ occurs in S. Example 4.18 shows that this property fails to entail $\mathscr{C}' \preceq \mathscr{C}$ (in other words, the converse of Proposition 4.2 fails).

Lemma 4.3 *Let S be a stripped sequence. For $1 \leq i < j \leq \sum_{e \in \mathbb{E}} \mu_{\mathscr{E}}(e)$, any occurrence of a chronicle $(\mathscr{E}, \mathcal{T})$ in S is also an occurrence of $(\mathscr{E}, \mathcal{T} \cup \{(e_i, i)[-\infty : +\infty](e_j, j)\})$ in S. The converse holds, too.*

Together with Proposition 4.2, Lemma 4.3 shows that adding/removing unlimited temporal constraints makes no change to the set of stripped sequences in which a given chronicle occurs. Careful, $(\mathscr{E}, \mathcal{T})$ and $(\mathscr{E}, \mathcal{T} \cup \{(e_i, i)[-\infty : +\infty](e_j, j)\})$ need not be equivalent: for $1 \leq i < j \leq \sum_{e \in \mathbb{E}} \mu_{\mathscr{E}}(e)$, if $(e_i, i)[-\infty : +\infty](e_j, j) \notin \mathcal{T}$ then $(\mathscr{E}, \mathcal{T}) \prec (\mathscr{E}, \mathcal{T} \cup \{(e_i, i)[-\infty : +\infty](e_j, j)\})$.

Example 4.19 (Chronicles with Repetition of Event Types)
     Here, we illustrate the impact of multiset order on the occurrences of chronicles in stripped sequences. We exhibit two cases of chronicle pairs with a repetition of an event type.

(continued)

Example 4.19 (continued)

The first case illustrates two chronicles that look very similar $\mathscr{C} = (\{\{ABCC\}\}, \{(A, 1)[1, 2](C, 3), (B, 2)[3, 4](C, 4)\})$ and $\mathscr{C} = (\{\{ABCC\}\}, \{(A, 1)[1, 2](C, 4), (B, 2)[3, 4](C, 3)\})$ (see Fig. 4.3). The only difference lays in the position of the C event type in the temporal constraints.

Let us now consider the stripped sequence $S = \langle (A, 1) \ (C, 2) \ (B, 3) \ (C, 7) \rangle$. According to Definition 4.2, $\mathscr{C}$ occurs in S ($\mathscr{C} \in S$), but $\mathscr{C}'$ does not occur in S. The second condition on the embedding enforces the embedding of a chronicle occurrence to satisfy the multiset order. In case of $\mathscr{C}'$, it is not possible to find an embedding that satisfy both the temporal constraints and the multiset order.

Semantically, the temporal constraints between A and C are interpreted differently. For $\mathscr{C}$, it states that *the first C in the occurrence must appear between 1 and 2 time units after A*. For $\mathscr{C}'$, it states that *the second C in the occurrence must appear between 1 and 2 time units after A*.

Symmetrically $\mathscr{C}'$ occurs in $S' = \langle (B, 1) \ (C, 5) \ (A, 6) \ (C, 7) \rangle$, but $\mathscr{C}$ does not occur in S'. None of these chronicles can occur in both S and S'. Hence, a chronicle can not handle a situation such as: "*C occurs between 1 and 2 time units after A and, in parallel, another C occurs between 3 and 4 time units after B*.". As we have seen, a chronicle requires to specify an order between occurrences of C.

Figure 4.4 gives two different chronicles which illustrate the interest of the order of multiset mapping to disambiguate chronicles. Intuitively, these two chronicles specify the same situation: two A separated of 1 to 2 time units. Technically, only the chronicle on the left occurs in sequence $\langle (A, 3) \ (A, 4) \rangle$. For the chronicle on the right, the temporal constraint enforces $(A, 2)$ to occur strictly before $(A, 1)$ (negative temporal constraints) and the multiset position enforces $(A, 2)$ to occur strictly after $(A, 1)$. Both are incompatible. Hence, the chronicle on the left is the only right way to specify the behavior described above.

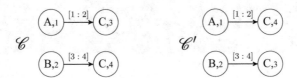

Fig. 4.3 $\mathscr{C}$ and $\mathscr{C}'$ look very similar but they do not occur in the same stripped sequences. The order of repeated event types plays a role in chronicle occurrences

Fig. 4.4 $\mathscr{C}$ and $\mathscr{C}'$ looks equivalent, but the chronicle on the right can not occur in any stripped sequence: the multiset order and the temporal constraints are incompatible

4.3   Counting Occurrences of a Chronicle

The previous section was devoted to the occurrence of a chronicle in a sequence. But, for some application, it is interesting to count how many times a chronicle occurs in a sequence. Intuitively, the count of a chronicle $\mathscr{C}$ that is more specific that another chronicle $\mathscr{C}'$ is equal or lower than the count of $\mathscr{C}'$.

In this section, we propose a method to count occurrences of a chronicle in a sequences that adheres to this intuition. This method is inspired by the notion of *maximal independent set* [74].

Definition 4.3   For a sequence S of length n and a chronicle $\mathscr{C}$ of size m, the count of all occurrences of the chronicle $\mathscr{C}$ in the sequence S, denoted $\mathsf{count-all}_S(\mathscr{C})$, is defined by:

$$\mathsf{count-all}_S(\mathscr{C}) = \left|\left\{ f \in [n]^m \,|\, f \text{ is an embedding of } \mathscr{C} \text{ in } S\right\}\right|$$

The Example 4.20 illustrates that the counting of all the occurrences of a chronicle is not anti-monotonic.

> *Example 4.20 (Chronicles with Repetition of Event Types)*
> Let S be a stripped sequence $\langle (A,1), (B,2), (B,3), (C,5)\rangle$
> Consider the chronicles $\mathscr{C} = (\mathcal{E}, \mathcal{T})$ and $\mathscr{C}' = (\mathcal{E}', \mathcal{T}')$ where $\mathcal{E} = \{\!\{AC\}\!\}$, $\mathcal{E}' = \{\!\{ABC\}\!\}$, $\mathcal{T} = \{(A, 1)[1 : 5](C, 2)\}$ and $\mathcal{T}' = \{(A, 1)[1 : 5](C, 3), (A, 1)[1 : 4](B, 2), (B, 2)[1 : 4](C, 3)\}$. It is clear that $\mathscr{C} \preceq \mathscr{C}'$.
> Figure 4.5 illustrates the occurrences of $\mathscr{C}$ and $\mathscr{C}'$ in S. There is only one occurrence of $\mathscr{C}$ in S, nonetheless there are two occurrences of $\mathscr{C}'$ in S. This is due to the multiple instances (2) of B in S. Each of these B yields an occurrence of $\mathscr{C}'$ sharing the same mappings for A and B with the occurrence of $\mathscr{C}$ (the temporal constraints are not restrictive in this case). Then, the number of occurrences of $\mathscr{C}'$ is twice the number of occurrences of $\mathscr{C}$.

In the seminal work of Mannila [54], the notion of *minimal occurrence* has been proposed for sequential patterns. It coincides with the idea that the overall period of the occurrence of one minimal occurrence can not be included in the period of another. His definition of minimal occurrence takes advantage of the strict order of the events in an occurrence imposed by the sequential pattern. With chronicle,

Fig. 4.5 More occurrences of a chronicle $\mathscr{C}'$ that is more specific than the chronicle $\mathscr{C}$. The dashed arrows illustrate the chronicle embeddings. The red and blue arrows illustrate the alternative embedding for $\mathscr{C}'$. This figure illustrates Example 4.20

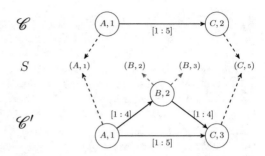

we can not guarantee that the last event of an occurrence will always match the same event type in the chronicle. For this reason, we were inspiration by the field of graph mining which also encounters the same kind of issue while counting the subgraph isomorphims in a graph [74]. The idea of Vanetik et al. [74] is to evaluate the edge-disjoint isomorphisms. The notion of *disjoint embedding* defined hereafter is inspired from this notion. Intuitively, it identifies *conflicting* embeddings when they use the same pairs of events in the sequence like in Example 4.20 in which the $((A, 1), (C, 5))$ pair appears in both embeddings. This leads us to define the number of occurrences of a chronicle as the maximum number of *non-conflicting* embeddings.

The definition below formally introduces these notions, and Example 4.21 illustrates them.

Definition 4.4 (Distinct and Disjoint Occurrences/Embeddings) Let $\mathscr{C} = (\mathcal{E}, \mathcal{T})$ be a chronicle of size m, S a stripped sequence and f and f' be two embeddings of $\mathscr{C}$ in S.

- f and f' are **distinct** iff there exists $i \in [m]$ such that $f(i) \neq f'(i)$.
- f and f' are **disjoint** iff

$$\{(f(i), f(j)) \mid (e_i, i)[l : u](e_j, j) \in \mathcal{T}\} \cap$$
$$\{(f'(i), f'(j)) \mid (e_i, i)[l : u](e_j, j) \in \mathcal{T}\} = \emptyset \qquad (4.1)$$

Definition 4.5 Let $\mathscr{C}$ be a chronicle of size m and S a sequence. $\mathsf{count}_S(\mathscr{C})$ is the cardinal of the largest set of disjoint occurrences of $\mathscr{C}$ in S.

Proposition 4.3 *Let S be a sequence, $\mathsf{count}_S(\cdot)$ is anti-monotonic wrt $\preceq$.*

Example 4.21 (Disjoint Occurrences in a Sequence) Let S be a stripped sequence $\langle (A,1), (B,2), (A,3), (B,4), (C,5) \rangle$.

(continued)

Example 4.21 (continued)

Consider the chronicles $\mathscr{C}_1 = (\mathcal{E}, \mathcal{T})$, $\mathscr{C}_2 = (\mathcal{E}', \mathcal{T}_2)$, $\mathscr{C}_3 = (\mathcal{E}', \mathcal{T}_3)$ and $\mathscr{C}_4 = (\mathcal{E}', \mathcal{T}_4)$ where $\mathcal{E} = \{\{AB\}\}$, $\mathcal{E}' = \{\{ABC\}\}$, $\mathcal{T}_1 = \{(A, 1)[-1 : 5](B, 2)\}$, $\mathcal{T}_2 = \{(A, 1)[-1 : 5](B, 2), (A, 1)[0 : 5](C, 3)\}$, $\mathcal{T}_3 = \{(A, 1)[0 : 5](C, 3), (A, 1)[-1 : 5](B, 2), (B, 2)[0 : 5](C, 3)\}$ and $\mathcal{T}_4 = \{(A, 1)[0 : 5](C, 3), (A, 1)[-1 : 5](B, 2), (B, 2)[2 : 5](C, 3)\}$. These four chronicles are illustrated in Fig. 4.6.

It is clear that $\mathscr{C}_1 \preceq \mathscr{C}_2 \preceq \mathscr{C}_3 \preceq \mathscr{C}_4$.

The count of occurrences of chronicles $\mathscr{C}_1$ to $\mathscr{C}_4$ are respectively: 4, 4, 2 and 1.

The occurrences of $\mathscr{C}_1$ are: $\langle (A,1), (B,2) \rangle$, $\langle (A,1), (B,4) \rangle$, $\langle ((B,2), (A,3) \rangle$ and $\langle (A,3), (B,4) \rangle$. For this chronicle, all the embeddings are disjoints. Then, it is the maximal set of occurrences.

The four occurrences of $\mathscr{C}_2$ are: $\langle (A,1), (B,2), (C,5) \rangle$, $\langle (A,1), (B,4), (C,5) \rangle$, $\langle ((B,2), (A,3), (C,5) \rangle$ and $\langle (A,3), (B,4), (C,5) \rangle$. In this case, we can notice that the pair $((B,4), (C,5))$ is used in several occurrences. According to our definition of disjoint occurrences and since there is no temporal constraints between $(B, 2)$ and $(C, 3)$, these occurrences are disjoint.

Chronicle $\mathscr{C}_3$ having an additional temporal constraint between $(B, 2)$ and $(C, 3)$, two of the occurrences of $\mathscr{C}_2$ are excluded in the occurrences of $\mathscr{C}_3$. Indeed, the fact that there are only two B and one C in the sequence leads to have at most two occurrences. Then, the occurrences of $\mathscr{C}_3$ are: $\langle (A,1), (B,2), (C,5) \rangle$ and $\langle (A,3), (B,4), (C,5) \rangle$. We can notice that the maximal set of disjoint occurrences is not unique. For instance, $\{\langle (A,1), (B,4), (C,5) \rangle$, $\langle (A,2), (B,3), (C,5) \rangle\}$ is another set of disjoint occurrences.

Finally, the only two occurrences of $\mathscr{C}_4$ are $\langle (A,3), (B,4), (C,5) \rangle$ and $\langle (A,1), (B,4), (C,5) \rangle$, but they are not disjoint. Then the count of its occurrences is 1.

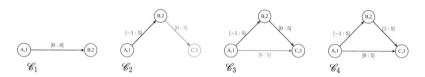

Fig. 4.6 Illustration of the four chronicles of Example 4.21. The red parts indicate the changes with respect to the previous chronicle

4.4 Conclusion

In this chapter, we focused on the occurrences of a chronicle. This task answers the problem of situation recognition. We introduced the notion of occurrence of a chronicle in a sequence and we analysed this definition in the light of our formal account of chronicles. The main result we obtained is related to the pre-order on the set of chronicles. We have the intuitive result that if a chronicle occurs in a sequence, then all its more specific chronicles also occur in the same sequence.

We also investigate the multiple occurrences of a chronicle in a sequence. Counting the multiple occurrences of a chronicle may be interesting in the context of the analysis of a long stream of events (for instance, to count the number of faulty behaviours described by a chronicle). For this specific task, we highlight that counting all occurrences of a chronicle fails to give an anti-monotonic measure. Then, we proposed the new measure that is anti-monotonic: the size of the largest set of disjoint occurrences.

Chapter 5
Inducing Chronicles from Event Sequences

Abstract In this chapter, we address the problem of inducing chronicles from a sequence or a collection of temporal sequences. This important problem is motivated by the need to identifies pieces of information that are recurrent in temporal sequence. This chapter show that a chronicle can be used to abstract the information from several sequences. We start by showing that a sequence can be represented by a chronicle (up to a time translation) to bridge the space of sequences and the space of chronicles. Then, we address the problem of summarizing a collection of sequences by a unique chronicle. This abstraction of a collection of sequences benefits from the results on the spaces of chronicles, and more precisely on the semi-lattice space of pairwise flush chronicle which allow to identify an abstraction as a kind of least general generalisation of the sequences. Finally, we focus on mining frequent chronicles from a collection of sequences using the framework of formal concept analysis.

Keywords Temporal abstraction · Chronicle mining · Formal concept analysis

5.1 Introduction

From now, we only handle how a chronicle occurs in a sequence. In this chapter, we address the opposite problem of inducing chronicles from event sequences. The objective is twofold: construct an abstract representation of a collection of sequences that summarize well a chronicle, or the induction of a collection of chronicles occurring with a certain interest in the collection of sequences.

5.2 Generating a Chronicle from a Sequence

From now, we mainly deal with how a chronicle occurs in a stripped sequence. The aim of this section is to define a chronicle that, in a sense, is represented by a stripped sequence. Such a chronicle amounts to the most specific (it is unique up to a time translation) chronicle that occurs in the stripped sequence.

Definition 5.1 Let $S = \langle (e_1, t_1), \ldots, (e_n, t_n) \rangle$ be a stripped sequence. Then, $\delta(S) = (\mathcal{E}_{\delta(S)}, \mathcal{T}_{\delta(S)})$ is the chronicle such that

- $\mathcal{E}_{\delta(S)} = \{\!\{ e_{\varphi(1)}, \ldots, e_{\varphi(n)} \}\!\}$
- $\mathcal{T}_{\delta(S)} = \{ (e_{\varphi(i)}, i)[t : t](e_{\varphi(j)}, j) \mid t_{\varphi(j)} - t_{\varphi(i)} = t, \ 1 \leq i < j \leq n \}.$

where $\varphi : [n] \to [n]$ is the permutation over $[n]$ such that $e_{\varphi(i)} \leq_{\mathbb{E}} e_{\varphi(i+1)}$ and if $e_{\varphi(i)} = e_{\varphi(i+1)}$ then $\varphi(i) < \varphi(i+1)$.

> *Example 5.22 (Chronicle Generated from a Stripped Sequence)* A graphical representation of the chronicle generated from the stripped sequence $\langle (A, 10), (B, 12), (C, 13), (A, 15) \rangle$ is depicted in Fig. 5.1. In this example, the images of 1, 2, 3 and 4 by φ are respectively 1, 3, 4 and 2. Please note that the chronicle encodes only the inter-event durations. Knowing the timestamps of at least event is required to reconstruct the whole timestamps of the sequence. As a consequence, the same chronicle is generated from a translated sequence, for instance $\langle (A, 4), (B, 6), (C, 7), (A, 9) \rangle$.

Lemma 5.1 φ in Definition 5.1 is unique.

There can be several permutations φ satisfying $e_{\varphi(i)} \leq_{\mathbb{E}} e_{\varphi(i+1)}$ but they all give the same *sequence* enumerating $\{\!\{ e_1, \ldots, e_n \}\!\}$. However, they would not always give *exactly* the same $\mathcal{T}_{\delta(S)}$ and this is the reason for the extra condition in Definition 5.1: if $e_{\varphi(i)} = e_{\varphi(i+1)}$ then $\varphi(i) < \varphi(i+1)$. This is illustrated in Example 5.23.

Fig. 5.1 The chronicle $\delta(S)$ generated from the stripped sequence $S = \langle (A, 10), (B, 12), (C, 13), (A, 15) \rangle$

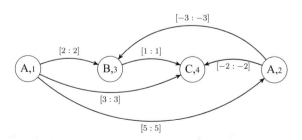

Example 5.23 (A Unique Event Permutation Satisfies Conditions of Definition 5.1) An illustration is as follows.

$$S = \cdots (A, t_4), (B, t_5), (A, t_6) \cdots$$
$$\mathcal{E}_{\delta(S)} = AAB \cdots$$

The enumeration $AAB \cdots$ can be generated (from S) by two permutations φ_1 and φ_2 described next:

$$\varphi_1 : \begin{array}{ll} 1 \mapsto 4 & A \mapsto (A, t_4) \\ 2 \mapsto 6 & A \mapsto (A, t_6) \\ 3 \mapsto 5 & B \mapsto (B, t_5) \\ \vdots & \vdots \end{array}$$

Calculating $\mathcal{T}_{\delta(S)}$ then gives

$$
\begin{array}{lll}
i = 1 \quad j = 3 & \rightsquigarrow & (e_4, 1)[t_5 - t_4](e_5, 3) \quad \text{i.e.} \quad (A, 1)[t_5 - t_4](B, 3) \\
i = 2 \quad j = 3 & \rightsquigarrow & (e_6, 2)[t_5 - t_6](e_5, 3) \quad \text{i.e.} \quad (A, 2)[t_5 - t_6](B, 3) \\
i = 1 \quad j = 2 & \rightsquigarrow & (e_4, 1)[t_6 - t_4](e_6, 2) \quad \text{i.e.} \quad (A, 1)[t_6 - t_4](A, 2)
\end{array}
$$

Now, the second permutation φ_2 is:

$$\varphi_2 : \begin{array}{ll} 1 \mapsto 6 & A \mapsto (A, t_6) \\ 2 \mapsto 4 & A \mapsto (A, t_4) \\ 3 \mapsto 5 & B \mapsto (B, t_5) \\ \vdots & \vdots \end{array}$$

Calculating $\mathcal{T}_{\delta(S)}$ would give

$$
\begin{array}{lll}
i = 1 \quad j = 3 & \rightsquigarrow & (e_6, 1)[t_5 - t_6](e_5, 3) \quad \text{i.e.} \quad (A, 1)[t_5 - t_6](B, 3) \\
i = 2 \quad j = 3 & \rightsquigarrow & (e_4, 2)[t_5 - t_4](e_5, 3) \quad \text{i.e.} \quad (A, 2)[t_5 - t_4](B, 3) \\
i = 1 \quad j = 2 & \rightsquigarrow & (e_6, 1)[t_4 - t_6](e_4, 2) \quad \text{i.e.} \quad (A, 1)[t_4 - t_6](A, 2)
\end{array}
$$

Please observe that φ_2 violates the extra condition in Definition 5.1, and that is: if $e_{\varphi(i)} = e_{\varphi(i+1)}$ then $\varphi(i) < \varphi(i + 1)$. Indeed, $e_6 = e_{\varphi_2(1)} = e_{\varphi_2(2)} = e_4$ but $6 = \varphi_2(1) > \varphi_2(2) = 4$.

We now give some results about the chronicle generation from a stripped sequence. Taken together, the next two lemmas prove that $\delta(S)$ is a finite slim chronicle.

Lemma 5.2 $\mathcal{T}_{\delta(S)}$ *is finite.*   ☐

Lemma 5.3 $\delta(S)$ *is a slim chronicle.*

In addition, we have also that $\delta(S)$ is a simple chronicle.

Lemma 5.4 *For S a stripped sequence, $\delta(S)$ is a simple chronicle.*   ☐

The following two results establish that $\delta(S)$ is the most specific chronicle that occurs in S.

Proposition 5.1 $\delta(S)$ *occurs in S.*

Proposition 5.2 $\mathscr{C} \preceq \delta(S)$ *iff $\mathscr{C}$ occurs in S.*

Let us know discuss the choice of the definition of a chronicle generation from a stripped sequence. Indeed, there are two appealing alternative definitions for that.

A first alternative set of temporal constraints generated from a chronicle $\delta'(S)$ from a stripped sequence would be

$$\mathcal{T}_{\delta'(S)} = \left\{ (e_{\varphi(i)}, i)[t : t](e_{\varphi(i+1)}, i + 1) \mid t_{\varphi(i+1)} - t_{\varphi(i)} = t, \ 1 \leq i < n \right\}.$$

A second alternative would be

$$\mathcal{T}_{\delta''(S)} = \left\{ (e_{\alpha(i)}, \varphi^{-1}(\alpha(i)))[t : t](e_{\beta(i)}, \varphi^{-1}(\beta(i))) \mid t_{i+1} - t_i = t, \ 1 \leq i < n \right\}$$

where α and β give the appropriate direction of the temporal constraint according to Lemma 2.1

$$(\alpha(i), \beta(i)) = \begin{cases} (i, i + 1) \text{ if } e_i \leq_{\mathbb{E}} e_{i+1} \\ (i + 1, i) \text{ otherwise.} \end{cases}$$

These two encodings of a stripped sequence determine some (non-unique) minimal chain chronicles. They are minimal in the sense that removing any edge (i.e., temporal constraint) from the chronicle fails to give a chronicle characterizing, up to a time translation, the stripped sequence.

A problem with this definition is precisely that there are non-unique minimals (the Example 5.24 illustrates two different chronicles yielded from the same stripped sequence). There is no reason to select one instead of another. The definition we adopted avoids this choice by generating all the temporal constraints between each pair of chronicles. As shown in the example, this is the condition under which Proposition 5.2 holds.

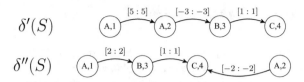

Fig. 5.2 Alternative chronicles generated from $S = \langle (A, 10), (B, 12), (C, 13), (A, 15) \rangle$

Example 5.24 (Alternative Generations of a Chronicle from a Sequence)
This example illustrates the two alternative generations of a chronicle (see
the text above) from the sequence of the Example 5.22. Figure 5.2 graphically
represents these chronicles.

The first encoding graphically maintains the event type order such that
the temporal constraints are between each consecutive pairs of events. The
images of the function ϕ of δ' are still 1, 3, 4 and 2 (similar as δ). The second
encoding graphically maintains the original temporal order of the events in
the stripped sequence. The images of pairs of events by the functions (α, β)
in the example are respectively $(1, 2)$, $(2, 3)$ and $(4, 3)$ (but not $(3, 4)$ because
$e_4 <_\mathbb{E} e_3$). Thus, the last temporal constraint is in the reverse time direction.

We call these "chain chronicles" as each of them can be graphically represented by a chain of events with a temporal constraint over two consecutive
events (see Fig. 5.2).

We have that $\delta'(S) \preceq \delta(S)$ and $\delta''(S) \preceq \delta(S)$ (Lemma 2.7).

Let us now consider the chronicle $\mathscr{C} = (\mathcal{E}, \mathcal{T})$, where $\mathcal{E} = \{\!\{AC\}\!\}$ and
$\mathcal{T} = \{(A, 1)[0 : 3](C, 2)\}$. Then, it is easy to check that $\mathscr{C}$ occurs in S.

Nonetheless, $\mathscr{C} \npreceq \delta'(S)$. The two possible multiset embeddings from $\mathcal{E}_{\delta'(S)}$
to $\mathcal{E}$ are $\{1 \mapsto 1, 2 \mapsto 4\}$ and $\{1 \mapsto 2, 2 \mapsto 4\}$, but there is no temporal
constraint in $\mathcal{T}_{\delta'(S)}$ in between $(A, 1)$ and $(C, 4)$ nor in between $(A, 2)$ and
$(C, 4)$.

Similarly, $\mathscr{C} \npreceq \delta''(S)$. The same multiset embeddings apply. Again,
there is no temporal constraint in $\mathcal{T}_{\delta'(S)}$ in between $(A, 1)$ and $(C, 4)$. This
discard the first embedding. Considering the second embedding, the temporal
constraint $(A, 2)[-2, -2](C, 4) \npreceq (A, 1)[0, 3](C, 2)$, then it is also to discard
and $\mathscr{C}$ is not less specific than $\delta''(S)$.

5.3   Inducing a Chronicle from Event Sequences

In this section chapter, we address the problem of inducing a single chronicle from
a sequence in a collection of temporal sequences, so-called a dataset of sequence in
the following.

5.3.1   Dataset of Event Sequences

Let us first remind that we previously defined the (temporal) events being a pair (e, t) where $e \in \mathbb{E}$ is an event type and $t \in \mathbb{T}$ is a timestamp (see Definition 2.1). In Definition 4.1, we introduced the sequence as a pair made of an identifier and a stripped sequence, i.e. an ordered set of events. In this chapter, we deal with a dataset of event sequences (or dataset of sequences for short), and we investigate how chronicle may abstract the sequences at hand.

Definition 5.2 A **dataset of sequences** is a finite unordered set of sequences.

Notation For $\mathcal{D}$ a dataset of sequences, let $Str(\mathcal{D})$ denote the set of stripped sequences to be found in $\mathcal{D}$ (formally, $S \in Str(\mathcal{D})$ iff $\langle I, S \rangle \in \mathcal{D}$ for some I). There could be multiple copies of the same stripped sequence S in $\mathcal{D}$, each with a different identifier I, but $Str(\mathcal{D})$ is to contain only one instance of S.

Fact 2 For $\mathcal{D}$ a dataset of sequences, $\left(2^{Str(\mathcal{D})}, \subseteq\right)$ is a poset. $\qquad\qquad\square$

Example 5.25 (Sequence Dataset) Table 5.1 illustrates a dataset of sequences whose corresponding chronicles, denoted $\mathscr{C}_1, \dots, \mathscr{C}_5$, and depicted in Fig. 5.3. Note that the labels of the sequences are not yet used in this chapter.

Table 5.1 The dataset of sequences $\mathcal{D}$ for Example 5.25

SID	Stripped sequence
1	$\langle (A, 1), (B, 4), (B, 5), (C, 6) \rangle$
2	$\langle (A, 3), (B, 7), (C, 8) \rangle$
3	$\langle (A, 2), (D, 4), (B, 6), (C, 7) \rangle$
4	$\langle (A, 2), (B, 4), (C, 5) \rangle$
5	$\langle (D, 1), (A, 4), (B, 4), (D, 5) \rangle$

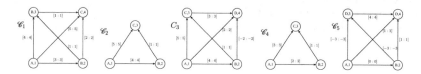

Fig. 5.3 Chronicles generated from the sequences of the dataset in Table 5.1

5.3.2   Abstracting a Dataset of Sequences with Chronicles

In this section, we give the principle of the abstraction of a set of sequences, i.e. a dataset of sequences, with chronicles. The motivation for searching such chronicles is to use the generalization capability of a chronicle to sum up a collection of sequences. We would like that an abstraction of sequences captures the common parts of the set of sequences. This means that an abstraction has to represent a small part in each sequence.

Intuitively, chronicles are interesting to generalize of a set of sequences. We expect the interval of temporal constraints can generalize various generalization. We have defined the occurrence of a chronicle in a sequence. Occurring in a sequence means that the chronicle recognizes a part of the sequence as a realization of the abstract behavior described in the chronicle. For a dataset of sequences, $\mathcal{D}$, we would like chronicles, e.g. $\mathcal{C}$ that occurs in all sequences $S \in \mathcal{D}$.

As we introduced in the previous section a way to represent a sequence by a chronicle, the problem is to find a chronicle $\mathcal{C}$ such that $\mathcal{C} \preceq \delta(S)$ for all $S \in \mathcal{D}$.

The more specific is the chronicle, the more meaningful is the abstraction. Indeed, the empty chronicle is a universal generalization of a dataset, but does not provide real insight about the dataset. For this reason, we are interested in the least general generalization of the set of chronicle generated from the sequence $\{\delta(S) \mid S \in \mathcal{D}\}$.

Our problem is that the formal account of chronicles highlights that the set of chronicles, or event the set of simple chronicles (like chronicles generated from sequences according to Lemma 5.4) is simply a poset. As a consequence, there is no unique least general generalization (see Example 3.10).

We identified some restrictive subspaces of chronicles, the sets of pairwise flush chronicles, upon which there is a lower semi-lattice structure. Nonetheless, these subspaces can not directly be used to represent the sequences of a dataset. Indeed, such subspace enforces the chronicles to conform exactly with a given profile, which is not necessarily the case for the chronicles in $\{\delta(S) \mid S \in \mathcal{D}\}$. In practice, it means that all the sequence have the same multiset of events. This is definitively too restrictive.

The following section solves the problem by exploiting the notion of pairwise flush chronicles after applying a transformation of the chronicle generated from the sequences that projects them into a common subspace of chronicles that conform with a profile.

5.3.3   A Dataset Chronicles Reduct

We start this section by introducing the notion of chronicle reduct conforming with a profile P.

Definition 5.3 Let P be a profile and $\mathscr{C} = (\mathcal{E}, \mathcal{T})$ a chronicle conforming with P. We write $\mathcal{E} = \{\!\{e'_1, \ldots, e'_m\}\!\}$ (we use the primed notation e'_i instead of e_i to avoid confusion in applications of the definition and lemmas for multiset embeddings).

The chronicle reduct corresponding to a profile P, denoted $\Lambda_P(\mathscr{C})$ is given by

$$\Lambda_P(\mathscr{C}) \stackrel{\text{def}}{=} (P, \widetilde{\Lambda_{\mathcal{E}}(\mathcal{T})}).$$

where

$$\Lambda_{\mathcal{E}}(\mathcal{T}) \stackrel{\text{def}}{=} \{(e_{\lambda_{\mathcal{E}}(i)}, \lambda_{\mathcal{E}}(i))[l:u](e_{\lambda_{\mathcal{E}}(j)}, \lambda_{\mathcal{E}}(j)) \mid (e'_i, i)[l:u](e'_j, j) \in \mathcal{T}\}$$

$$\cup \{(e_{\lambda_{\mathcal{E}}(i)}, \lambda_{\mathcal{E}}(i))[-\infty:+\infty](e_{\lambda_{\mathcal{E}}(j)}, \lambda_{\mathcal{E}}(j)) \mid 1 \le i < j \le m\}$$

and for $i = 1, \ldots, m$,

$$\lambda_{\mathcal{E}}(i) \stackrel{\text{def}}{=} i + \sum_{e <_{\mathbb{E}} e'_i} \mu_P(e) - \sum_{e <_{\mathbb{E}} e'_i} \mu_{\mathcal{E}}(e).$$

Remark In the definition of $\Lambda_{\mathcal{E}}(\mathcal{T})$, the second set adds unlimited temporal constraints between each pair of temporal events of $\mathcal{E}$. Nonetheless, the slimming operation applied on this set does remove all of these unlimited temporal constraints for which there was already a temporal constraint in $\mathscr{C}$.

Example 5.27 illustrates the construction of a chronicle reduct corresponding to a profile P.

Lemma 5.5 $\Lambda_P(\mathscr{C})$ *is a slim chronicle.* □

Remark In Definition 5.3, the chronicle $\mathscr{C}$ does not necessarily conform *strictly* with P. A chronicle reduct can hold strictly more event types than its chronicle. Especially, it means that $\mathcal{E} \in P$.

Lemma 5.6 $\lambda_{\mathcal{E}}$ *is a multiset embedding for* $\mathcal{E} \in P$. □

Let us introduce the notion of *least-embedding* to characterize the $\lambda_{\mathcal{E}}$ embedding.

Notation Let $\mathcal{E}$ and $\mathcal{E}'$ be two multisets such that $\mathcal{E} \in \mathcal{E}'$, the least embedding of $\mathcal{E}'$ in $\mathcal{E}$ is the embedding defined by:

$$\lambda_{\mathcal{E} \to \mathcal{E}'}(i) = i + \sum_{e <_{\mathbb{E}} e'_i} \mu_{\mathcal{E}'}(e) - \sum_{e <_{\mathbb{E}} e'_i} \mu_{\mathcal{E}}(e), \; \forall i \in [n].$$

Example 5.26 (Least-Embeddings) Let $\mathcal{E} = \{\!\{BBC\}\!\}$ and $\mathcal{E}' = \{\!\{ABBBC\}\!\}$, whose multiplicity functions are

$$\mu_\mathcal{E}(A) = 0 \quad \mu_{\mathcal{E}'}(A) = 1$$
$$\mu_\mathcal{E}(B) = 2 \quad \mu_{\mathcal{E}'}(B) = 3$$
$$\mu_\mathcal{E}(C) = 1 \quad \mu_{\mathcal{E}'}(C) = 1$$

Then, the least embedding is obtained by

$$\lambda_{\mathcal{E}\to\mathcal{E}'}(1) = 1 + \mu_{\mathcal{E}'}(A)$$
$$= 2$$
$$\lambda_{\mathcal{E}\to\mathcal{E}'}(2) = 2 + \mu_{\mathcal{E}'}(A) - \mu_\mathcal{E}(A)$$
$$= 2 + 1 - 0$$
$$= 3$$
$$\lambda_{\mathcal{E}\to\mathcal{E}'}(3) = 3 + \mu_{\mathcal{E}'}(A) + \mu_{\mathcal{E}'}(B) - \mu_\mathcal{E}(A) - \mu_\mathcal{E}(B)$$
$$= 3 + 1 + 3 - 0 - 2$$
$$= 5$$

Figure 5.4 illustrates this embedding.

The intuition behind the mapping of a chronicle reduct is to take advantage of the distinction between the case that (e_i, i) and (e_j, j) have no temporal constraint between them and the case that they are only related through an unlimited temporal constraint. Roughly speaking, when mapping $\mathcal{C}$ to a chronicle with profile P_D (intuitively, the less specific profile compatible with all chronicles induced by the sequences in $\mathcal{D}$), no extra copy of e in P_D is to be involved in a temporal constraint of the resulting chronicle. In contrast, for (e_i', i) and (e_j', j) both in $\mathcal{E}$ (with, say, $i < j$), the unlimited temporal constraint $(e_i', i)[-\infty : +\infty](e_j', j)$ is to be added

Fig. 5.4 Illustration of least-embedding of the multiset $\mathcal{E} = \{\!\{BBC\}\!\}$ into the multiset $\mathcal{E}' = \{\!\{ABBBC\}\!\}$. The first instance of an event type in $\mathcal{E}$ is mapped to the first instance of the same event type in $\mathcal{E}'$. See the calculation details in Example 5.26

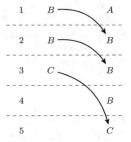

unless there is already in $\mathcal{T}$ a temporal constraint of the form $(e'_i, i)[l : u](e'_j, j)$ (in the latter case, the addition would make the chronicle to fail to be slim).

This intuition leads us to define a notion of pure chronicle and to Lemma 5.7.

Definition 5.4 A pure chronicle $\mathscr{C} = (\mathcal{E}, \mathcal{T})$ is such that all the following conditions hold:

- $\mathscr{C}$ is finite
- $\mathscr{C}$ is slim
- $\sum_{e \in \mathbb{E}} \mu_{\mathcal{E}}(e) > 1$
- $\mathcal{T}$ has no temporal constraint of the form $(e_i, i)[-\infty : +\infty](e_j, j)$.

The next to last condition in the definition means that $\mathcal{E}$ is neither empty nor a singleton set.

Lemma 5.7 *If the domain of Λ_P is restricted to the pure chronicles defined over P and $\mathbb{T}$, Λ_P is injective.*

Remark If we consider only chronicles having no unlimited temporal constraint, there is no loss of generality because Lemma 4.3 shows that adding/removing unlimited temporal constraints does not change the set of occurrences of a chronicle.

Example 5.27 (Chronicle Reduct Rewriting) Figure 5.5 illustrates adding vertices and arcs to a chronicle as just described. Assume $P = \{\!\{ABCDD\}\!\}$. Each chronicle reduct is a chronicle of which the multiset is P. Please observe that there is one additional node, namely $(D, 5)$ in the reduct at the top of the figure, and one additional node, namely $(C, 3)$, in the reduct at the bottom of the figure. These nodes can be distinguished from the others as they are involved in no temporal constraint.

For the chronicle at the top, $\lambda_{\mathcal{E}}$ is identity and for the chronicle at the bottom, it is defined by:

$$\lambda_{\mathcal{E}} : \begin{array}{l} 1 \mapsto 1 \\ 2 \mapsto 2 \\ 3 \mapsto 4 \\ 4 \mapsto 5 \end{array}$$

The plain arrows in the graphical representation of the chronicle reduct illustrate the temporal constraints of the original chronicle. The dashed arrows are unlimited temporal constraints added to supplement the temporal constraints between events present in the original chronicle.

Lemma 5.8 *Let $S = \langle (e_1, t_1), \ldots, (e_m, t_m) \rangle$ be a sequence conforming with a bounded profile P, then the chronicle $\Lambda_P(\delta(S)) \stackrel{\text{def}}{=} (P, \mathcal{T})$ is such that*

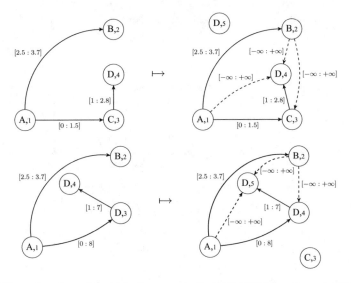

Fig. 5.5 Illustration of two chronicle reducts to a profile $\{\!\{ABCDD\}\!\}$ (see Example 5.27). In both cases, the chronicle on the right is a reduct of the chronicle on the left

$$\mathcal{T} = \{(e_{\lambda(i)}, \lambda(i)[d : d](e_{\lambda(j)}, \lambda(j)) \mid 1 \leq i < j \leq m\}$$

and λ selects a subset of event types in P.

Lemma 5.8 highlights the specific case of a chronicle generated from a sequence. The reduct of such a chronicle has a particular shape: there is a core multiset of event types (defined by λ) that are completely and simply connected, i.e., there is a unique temporal constraint between each pair of event types of this multiset, and the other event types are not involved in any temporal constraint. In addition, the interval in these temporal constraints is degenerate, i.e. both of its endpoinds are equal (d). Figure 5.6 of Example 5.28 illustrates such a particular shape of chronicles.

Lemma 5.9 *Let S be a sequence and P a profile such that $\delta(S)$ conforms with P, then $\Lambda_P(\delta(S))$ is a pure chronicle.*   □

Let us now come back to the problem of abstracting the dataset of sequences by a chronicle. We consider a reduct corresponding to the profile $\mathcal{E}_\mathcal{D}$ (remember, $\mathcal{D}$ is finite) whose multiplicity function is[1]

$$\mu_{\mathcal{E}_\mathcal{D}} \stackrel{\text{def}}{=} \max\{\mu_{\mathcal{E}_{\delta(S)}} \mid S \in Str(\mathcal{D})\}.$$

Intuitively, $\mathcal{E}_\mathcal{D}$ is the multiset consisting of as many copies of e, for each $e \in \mathbb{E}$, as the largest number of times e appears in $\delta(S)$, for S ranging over $Str(\mathcal{D})$.

---

[1] Reminder: $\mathcal{E}_{\delta(S)}$ is introduced in Definition 5.1.

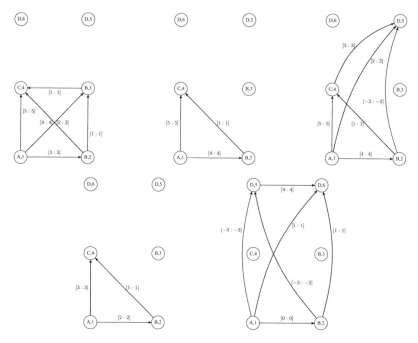

Fig. 5.6 Reducts of the dataset' chronicles (see Fig. 5.3) with the dataset profile $\{\!\{ABBCDD\}\!\}$ (see Example 5.28)

Lemma 5.10 *For all* $S \in \mathcal{D}$, $\delta(S)$ *conforms with* $\mathcal{E}_\mathcal{D}$. $\square$

Notation We use the abbreviation $\mathbb{E}_\mathcal{D}$ for the carrier of $\mathcal{E}_\mathcal{D}$, that is, $\mathbb{E}_\mathcal{D} = \{e \in \mathbb{E} \mid \mu_{\mathcal{E}_\mathcal{D}}(e) \neq 0\}$.

Intuitively, the induced chronicles are to be rewritten through their reduct corresponding to the profile $\mathcal{E}_\mathcal{D}$. We denote $\Lambda_P(\mathcal{D}) = \{\Lambda_P(\delta(S)) \mid S \in \mathcal{D}\}$ the set of chronicle reducts of the dataset $\mathcal{D}$.

Lemma 5.11 *If* $\mathscr{C} \in \Lambda_{\mathcal{E}_\mathcal{D}}(\mathcal{D})$, *then* $\mathscr{C}$ *is a pure chronicle.* $\square$

Lemma 5.11 indicates that the use of the dataset profile to reduct a chronicle generated from a sequence of the dataset leads to chronicles without unlimited temporal constraints.

We can now propose the construction of a chronicle that abstract the dataset of sequences.

The rewriting reduces the structure over the original chronicles to a collection of chronicles that are pairwise flush. All the elements of $\Lambda_{\mathcal{E}_\mathcal{D}}(\mathcal{D})$ are included in $\widetilde{\mathcal{C}_{f\mathcal{E}_\mathcal{D}}}$ the space of the finite slim chronicles that conform exactly with the profile $\mathcal{E}_\mathcal{D}$ — which is a lower semi-lattice according to Proposition 3.3. Then, the lgg (that is, glb for $\preceq$) of the chronicles of $\Lambda_{\mathcal{E}_\mathcal{D}}(\mathcal{D})$ is an abstraction of the set of sequences.

Notation glb $\Lambda_{\mathcal{E}_\mathcal{D}}(\mathcal{D})$ denotes the glb of the set of chronicles $\Lambda_{\mathcal{E}_\mathcal{D}}(\mathcal{D})$.

Lemma 5.12 *Let P be a profile and S be a sequence such that $\delta(S)$ conforms with P. Then, $\Lambda_P\,(\delta\,(S))$ is simple pure.*

Lemma 5.13 *Let $\mathscr{C} = (\mathcal{E}'', \mathcal{T})$ and $\mathscr{C}' = (\mathcal{E}'', \mathcal{T}')$ be two simple pure chronicles sharing the same multiset $\mathcal{E}''$, then $\mathscr{C} \curlywedge \mathscr{C}'$ is simple pure.*

Lemma 5.14 glb $\Lambda_{\mathcal{E}_\mathcal{D}}(\mathcal{D})$ *is a simple pure chronicle.*

Lemma 5.15 glb $\Lambda_{\mathcal{E}_\mathcal{D}}(\mathcal{D}) \preceq \Lambda_{\mathcal{E}_\mathcal{D}}(\delta(S))$ *for all $S \in \mathcal{D}$.* $\qquad\qquad\square$

Remark It is worth noting that glb $\Lambda_{\mathcal{E}_\mathcal{D}}(\mathcal{D})$ need not occur in a sequence of $\mathcal{D}$.

We recall that our objective is to find an abstraction for a collection of sequences. glb $\Lambda_{\mathcal{E}_\mathcal{D}}(\mathcal{D})$ seems to be a good candidate as it is the unique chronicle that meets the reduct of each chronicle generated from the sequences. Nonetheless, the above remark highlights an undesired behavior of a chronicle: Intuitively, the abstraction of a collection of sequences should occur in each of the sequences.

The following definition proposes a notion of abstraction derived from the minimal construction involved in a chronicle reduct.

Definition 5.5 For $\mathscr{C} = (\mathcal{E} = \{\!\{e_1, \ldots, e_n\}\!\}, \mathcal{T})$,

$$\overline{\mathscr{C}} \stackrel{\text{def}}{=} (\overline{\mathcal{E}}, \overline{\mathcal{T}})$$

on the condition that there exists a multiset embedding λ for $\overline{\mathcal{E}}$ in $\mathcal{E}$ satisfying both

$$\overline{\mathcal{E}} = \{\!\{e_{\lambda(i)} \in \mathcal{E} \mid \nexists(e_{\lambda(i)}, \lambda(i))[l : u](e_{\lambda(j)}, \lambda(j)) \in \mathcal{T} \,\wedge$$
$$\nexists(e_{\lambda(j)}, \lambda(j))[l : u](e_{\lambda(i)}, \lambda(i)) \in \mathcal{T}, \; j \in [n]\}\!\}$$

and

$$\overline{\mathcal{T}} = \{(e_i', i)[l : u](e_j', j) \mid (e_{\lambda(i)}, \lambda(i))[l : u](e_{\lambda(j)}, \lambda(j)) \in \mathcal{T}\}.$$

Then, $abs(\mathcal{D}) = \overline{\text{glb } \Lambda_{\mathcal{E}_\mathcal{D}}(\mathcal{D})}$ is called the **abstraction** of $\mathcal{D}$.

The operator $\bar{\cdot}$ removes the event types of a chronicle that are not connected with a temporal constraint to any other event type in the chronicle.

Lemma 5.16 *There is a unique multiset embedding from $\overline{\mathcal{E}}$ to $\mathcal{E}$.*

Lemma 5.17 $\mathcal{E}_{abs(\mathcal{D})} \Subset \mathcal{E}_{\delta(S)}$, *for all $S \in \mathcal{D}$.*

Proposition 5.3 $abs(\mathcal{D}) \Subset S$, *for all $S \in \mathcal{D}$.*

Now, $abs(\mathcal{D})$ seems to have a lot of the desired properties to be a good intuitive abstraction of a set of sequences.

Fig. 5.7 On the left, greatest lower bound of the chronicle reducts from Fig. 5.6 and, on the right, abstraction of a dataset (see Example 5.28)

Example 5.28 (Chronicle Abstraction (Continuing Example 5.25))

Let us continue Example 5.25 that was meant to illustrate the construction of chronicles from a dataset of sequences. The profile for the dataset is $\mathcal{E}_\mathcal{D} = \{\!\{ABBCDD\}\!\}$.

The chronicle reducts of the dataset, $\Lambda_{\mathcal{E}_\mathcal{D}}(\mathcal{D})$, consist of of the chronicles appearing in Fig. 5.6. Note that none of these chronicles has unlimited temporal constraints (cf Lemma 5.11).

The chronicle in Fig. 5.7 illustrates the intersection of all the chronicles in Fig. 5.6. All of these chronicles share the same multiset (namely, $\mathcal{E}_\mathcal{D}$), hence the intersection has the same multiset. As regards temporal constraints in these chronicles, all the multiset embeddings required in Definition 3.4 happen to be identities. Then, we only keep the temporal constraints that are shared by all the chronicles, and abstract their interval by the convex hull of the corresponding intervals in the chronicles to intersect. In this example, we only keep the temporal constraint between $(A, 1)$ and $(B, 2)$, and its interval is $[0 : 4]$ (convex hull of the intervals $[3 : 3]$, $[4 : 4]$, $[4 : 4]$, $[2 : 2]$ and $[0 : 0]$.

Example 5.29 (Abstraction Is Responsive to Positions)

This example gives again an illustration of the generalization of a dataset. The objective of this example is to stress that position in the multiset is important in our notion of abstraction.

Let us consider the following dataset:

SID	Stripped sequence
1	$\langle (A, 12), (B, 14) \rangle$
2	$\langle (A, 12), (A, 13), (B, 14) \rangle$
3	$\langle (A, 3), (A, 12), (B, 14) \rangle$

(continued)

Example 5.29 (continued)

One can notice that $\langle (A, 12), (B, 14) \rangle$ is a subsequence for each of the three stripped sequences. One might expect that this subsequence forms a recurrent pattern to be mentioned in the least general generalization of the dataset.

However, as per the chronicle reduct, the abstraction of the dataset is the chronicle $(\{\{AB\}\}, \{(A, 1)[2 : 11](B, 2)\})$. The upper bound of the interval is unexpectedly large because the interval $[2 : 2]$ might seem sufficient.

The second and third sequences contain an additional instance of A such that $(A, 12)$ is the first instance of A in the second sequence, but the second instance of A in the third sequence (in red). In the construction of intersection between the chronicle reducts of the sequences, the temporal constraint that is shared by the three chronicle reducts are in between the first occurrence of A and the first occurrence of B in the sequence. So that the temporal constraint induced by the red occurrence of A is used in the abstraction.

One can notice that the green instance of A in the second sequence has no impact on the abstraction. Yet, if the red instance of A is removed, the abstraction becomes $(\{\{AB\}\}, \{(A, 1)[2 : 2](B, 2)\})$.

By this example, we see that the abstraction $(\{\{AB\}\}, \{(A, 1)[2 : 11](B, 2)\})$ has to be read as: "*the delay between the first instance of A and the first instance of B lies in the interval* $[2 : 11]$".

5.3.4 Summary

In this section, we briefly summarize the overall approach (depicted in Fig. 5.8) we propose for an abstraction of a collection of temporal sequences.

We have seen that the space of (simple/finite) chronicles does not enjoy a semi-lattice structure. The principle we adopt to fix this then amounts to map the sequences in the dataset into a unique space of flushed chronicles. For this, we first apply the δ operator to create a chronicle from a sequence. Second, we apply the Λ operator that embeds the chronicle in the space of chronicles conforming with the multiset that subsumes all the sequences in the dataset ($\mathcal{E}_\mathcal{D}$).

This space of chronicles being a lower semi-lattice, it is possible to define a greatest lower bound which abstracts the collection of sequences. However, this glb lies in the space of chronicles conforming with the multiset $\mathcal{E}_\mathcal{D}$, and thus it does not necessarily occur in the sequences of the dataset. For this reason, the abstraction is defined by removing the "useless" events in the glb. This chronicle is unique and occurs in each sequence of the dataset.

Fig. 5.8 Diagram of the
approach taken for defining
an abstraction of a set of
sequences. See the text for
details

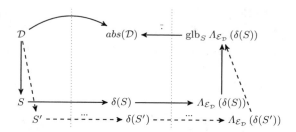

Remark Our definition of an abstraction makes a choice about the embedding to generalize among the embeddings between the multiset of the abstraction and a sequence. This choice is made by the Λ operator which embeds the temporal constraints between the events with the least indexes in $\mathcal{E}_{\mathcal{D}}$.

5.4 Mining Chronicles from Event Sequences

In this section, we address the task of mining chronicles from a collection of temporal sequences. This problem is an instantiation of the pattern mining task [42] to chronicles. We consider the task of frequent chronicles mining which aims at inducing chronicles that occur in at least σ sequences in the collection, where σ is defined by the user. Such chronicles are called *frequent chronicles*. Contrary to the previous section, in which we proposed to abstract the entire dataset by a unique chronicle, pattern mining extracts a collection of chronicles that occur in the dataset. The question arises whether chronicle mining can be linked to the notion of abstraction we have introduced. Indeed, a dual view of pattern mining is the identification of some subsets of sequences, with size larger than σ, from which some chronicle is *abstracted*.

Formal Concept Analysis (FCA) offers a framework to formalize this pattern mining task as a selection of some formal concept in a semilattice of concepts. FCA capture the duality of the problem between the space of chronicles and the space of the powerset of the sequence database. A concept represents both the subset of the dataset of sequences and its abstraction (a chronicle). The Sect. 5.4.3 reminds the connections between FCA and pattern mining in the work. It focuses on the work of Kuznetsov [47] on pattern structures. Since the early beginning of our work, our objective is to apply this framework to induce chronicles. If it applies, then we could benefit from decades of research on FCA-based pattern mining to design chronicle mining algorithms.

Applying this framework to induce chronicles requires a (semi-)lattice structure. The results of Chap. 3 and the notion of abstraction introduced in Sect. 5.3 are the building blocks for applying the FCA to chronicles.

Remark Designing a frequent chronicle mining algorithm does not require the framework of FCA. A frequent pattern mining algorithm simply requires two sufficient notions: (1) a poset of patterns, (2) a support measure which is anti-monotonic (with respect to the partial order on the poset).

Then, a frequent pattern mining algorithm consists in browsing the space of patterns and cutting the search space as soon as a pattern is not frequent (the anti-monotony of the support indicating that a more specific pattern cannot be frequent). The problem of pattern mining problem is then to optimize the browsing of the space in order to exhibit good algorithmic properties (such as completeness or time-efficiency). In this book, it is not our objective to investigate these questions and we refer the reader to the different strategies that have been proposed to extract frequent chronicles [2, 21, 27, 63].

5.4.1 Basic Notions

Notation For a sequence we write $\mathbf{S} = \langle I, \langle (e_1, t_1), (e_2, t_2), \ldots, (e_n, t_n) \rangle \rangle$. Please notice the difference in notation as we will use $\mathbf{S}$ to denote a sequence, but for the stripped sequence in it, we will use the non-bold symbol S.

Definition 5.6 Let $\mathcal{D}$ be a dataset of sequences and $\mathscr{C}$ a chronicle. The **support** of $\mathscr{C}$ in $\mathcal{D}$ is the number of sequences in $\mathcal{D}$ such that $\mathscr{C}$ occurs in S. In symbols,

$$supp_{\mathcal{D}}(\mathscr{C}) = |\{\mathbf{S} \in \mathcal{D} \mid \mathscr{C} \in S\}|.$$

Remark Please notice that the support counts the number of sequences, not the number of occurrences (i.e., of embeddings). Indeed, a sequence in which the chronicle has more than one embedding only counts once towards the support.

We refer the reader to Sect. 4.3 about counting the occurrences of a chronicle in a single sequence.

Proposition 5.4 *If $\mathscr{C}$ and $\mathscr{C}'$ are two chronicles such that $\mathscr{C}' \preceq \mathscr{C}$ then $supp_{\mathcal{D}}(\mathscr{C}') \geq supp_{\mathcal{D}}(\mathscr{C})$.*

The property of anti-monotonicity of the support in Proposition 5.4 holds on the set of chronicles. Accordingly, the support of the glb of $\mathscr{C}$ and $\mathscr{C}'$ cannot be less than the support of each of $\mathscr{C}$ and $\mathscr{C}'$.

Notation $\mathscr{C} \prec \mathscr{C}' \Leftrightarrow \mathscr{C} \preceq \mathscr{C}' \wedge \mathscr{C} \neq \mathscr{C}'$ for slim chronicles $\mathscr{C}$ and $\mathscr{C}'$.

Definition 5.7 A slim chronicle C is **closed** wrt a dataset of sequences $\mathcal{D}$ iff $supp_{\mathcal{D}}(\mathscr{C}) > supp_{\mathcal{D}}(\mathscr{C}')$ for every slim chronicle $\mathscr{C}'$ such that $\mathscr{C} \prec \mathscr{C}'$.

Table 5.2 The dataset of sequences $\mathcal{D}$ for Example 5.30

SID	Stripped sequence
1	$\langle (A, 1), (B, 4), (B, 5), (C, 6) \rangle$
2	$\langle (A, 3), (B, 7), (C, 8) \rangle$
3	$\langle (A, 2), (D, 4), (B, 6), (C, 7) \rangle$
4	$\langle (A, 2), (B, 4), (C, 5) \rangle$
5	$\langle (D, 1), (A, 4), (B, 4), (D, 5) \rangle$

Table 5.3 Chronicle supports in the dataset $\mathcal{D}$ (Table 5.2)

Chronicle	Support
$\mathcal{C}_1$	5
$\mathcal{C}_2$	3
$\mathcal{C}_3$	3
$\mathcal{C}_4$	2

Example 5.30 (Frequent Chronicles) Consider the following chronicles:

$$\mathcal{C}_1 = (\{\!\{A, B\}\!\}, \{(A, 1)[2:4](B, 2)\})$$

$$\mathcal{C}_2 = (\{\!\{A, B\}\!\}, \{(A, 1)[3:4](B, 2)\})$$

$$\mathcal{C}_3 = \left(\{\!\{A, B, C\}\!\}, \left\{ \begin{array}{l} (A, 1)[2:4](B, 2), \\ (B, 2)[1:1](C, 3) \end{array} \right\} \right)$$

$$\mathcal{C}_4 = (\{\!\{A, D\}\!\}, \{(A, 1)[-3:2](D, 2)\})$$

Considering the dataset of Table 5.2, the supports for each chronicle are given in Table 5.3.

Please observe that $\mathcal{C}_4$ has two embeddings in sequence 5 (similarly, $\mathcal{C}_1$ in sequence 1), but they are counted only once in the supports.

Also, $\mathcal{C}_1$ is closed.

5.4.2 Frequent Chronicles

In this section, we consider the frequent chronicles mining tasks. This task of chronicle induction was the very first one [27] and, at the time, the most addressed one in literature. Contrary to the abstraction of a complete dataset of sequences, frequent chronicle mining consists in abstracting sequences for different subsets of sequences. Then, a frequent chronicle is meant to capture regularities in a subset of the original dataset, but the set of frequent chronicles aims at capturing all the regularities in the dataset. In this sense, such a task is supposed to abstract the information contained in a dataset.

From now on, we assume a fixed user-defined threshold $\sigma \in \mathbb{N}$ has been set.

Definition 5.8 A chronicle $\mathscr{C}$ is **frequent** iff $supp_{\mathcal{D}}(\mathscr{C}) \geq \sigma$.

Corollary 5.1 *Let $\mathscr{C}$ and $\mathscr{C}'$ be two chronicles such that $\mathscr{C}' \preceq \mathscr{C}$, if $\mathscr{C}'$ is not frequent then $\mathscr{C}$ is not frequent.* $\square$

Corollary 5.1 of Proposition 5.4 states that being frequent is an anti-monotonic property over the poset of slim chronicles $(\mathcal{C}, \preceq)$. It arises from Proposition 5.4 which states the anti-monotonicity of the support measure. Obviously, it is also anti-monotonic on the poset of chronicles $(\widetilde{\mathcal{C}_f}, \preceq)$.

> *Example 5.31 (Frequent Chronicles (Continued from Example 5.30))* For $\sigma = 2$, $\mathscr{C}_1$, $\mathscr{C}_2$, $\mathscr{C}_3$ and $\mathscr{C}_4$ are frequent.
>
> $\mathscr{C}_1$, $\mathscr{C}_2$ and $\mathscr{C}_3$ are finite slim chronicles such that $\mathscr{C}_1 \preceq \mathscr{C}_2$ and $\mathscr{C}_1 \preceq \mathscr{C}_3$. In accordance with the anti-monotonicity of the support, $supp_{\mathcal{D}}(\mathscr{C}_1) \geq supp_{\mathcal{D}}(\mathscr{C}_2)$ and $supp_{\mathcal{D}}(\mathscr{C}_1) \geq supp_{\mathcal{D}}(\mathscr{C}_3)$.
>
> In accordance with the anti-monotonicity of being frequent:
>
> - for $\sigma = 6$, that $\mathscr{C}_1$ is not frequent entails that $\mathscr{C}_2$ and $\mathscr{C}_3$ are also not frequent,
> - for $\sigma = 3$, $\mathscr{C}_3$ is frequent and thus $\mathscr{C}_1$ is also frequent.

5.4.3 Formal Concept Analysis and Pattern Structures

In this Section, we recall some definitions and results from Ganter and Kuznetsov [32] who invented the notion of pattern structures —a general framework for frequent pattern mining algorithms which is inspired by formal concept analysis.

A *pattern structure* designates a specific kind of Galois connection between a powerset of examples and a set of patterns. In our case, patterns are chronicles and examples are stripped sequences.

We now provide some reminders in order to formally define patterns structures. The next section is dedicated to applying them for chronicles.

A **Galois connection** between two posets $\langle A, \leq_A \rangle$ and $\langle B, \leq_B \rangle$ is a pair of functions (φ, ψ) where $\varphi : A \to B$ and $\psi : B \to A$ such that $\varphi(a) \leq_B b$ iff $a \leq_A \psi(b)$.

An alternative definition of a Galois connection is a pair of antitone, i.e. order-reversing, functions $\varphi : A \to B$ and $\psi : B \to A$ between two posets A and B, such that

$$b \leq \varphi(a) \quad \Leftrightarrow \quad a \leq \psi(b) \tag{5.1}$$

We now recall the definition of a pattern structure due to Ganter and Kuznetsov [32].

Let G be some set, let $\underline{D} = (D, \sqcap, \sqsubseteq)$ be a meet-semilattice and let $\rho : G \to D$ be a mapping. Then $(G, \underline{D}, \rho)$ is called a **pattern structure**, provided that the set

$$\rho(G) \stackrel{\text{def}}{=} \{\delta(S) \mid S \in G\}$$

generates a complete subsemilattice $(D_\delta, \sqcap)$ of $\underline{D}$. This is satisfied in two situations: when $\underline{D}$ is complete, and when and G is finite.

Remark The original notation for the mapping function of a pattern structure is δ (instead of ρ). We introduce the notation ρ to avoid using the same notation as for the function to generate a chronicle from a stripped sequence (Definition 5.1).

Interpreting a pattern structure $(G, \underline{D}, \rho)$ in terms of pattern mining, G would correspond to the set of examples, D would be the set of patterns, with a Galois connection defined between the powerset of G and D according to the following operators:

$$\varphi(A) \stackrel{\text{def}}{=} \bigsqcap_{g \in A} \rho(g), \ A \subseteq G$$

and

$$\psi(d) \stackrel{\text{def}}{=} \{g \in G \mid d \sqsubseteq \rho(g)\}, \ d \in D.$$

Then, formal pattern concepts are defined as pairs of a subset of examples and a pattern that are determined by each other. For $g \subseteq G$, if $g = \psi\varphi(g)$ then g is closed under the closure operator on G and is said to be an *extent*, and a subset $d \sqsubset D$ with $d = \varphi\psi(d)$ is called an *intent*. A pair (g, d) is said to be a *pattern concept* with extent g and intent d. The system of all pattern concepts forms a complete lattice.

Remark The formal concept analysis framework introduces the notion of *formal concept* defined by the Galois connection between two powerset lattices (the powerset of examples and the powerset of descriptors). In a pattern structure, the lattice of the descriptions is a lattice of patterns. The natural relation on a powerset (subset relation) is replaced by an abstract relation specific to a domain of patterns.

Formal concept analysis and pattern structure analysis are strongly related to the problem of frequent pattern mining. The task of mining frequent patterns is described by Stumme [69] as follows: Given a set of objects G, a set of attributes D, a binary relation $I \subseteq G \times D$ (where $(g, d) \in I$ is read as "object g has attribute d"), and a threshold $minsupp \in \mathbf{N}$, determine all subsets X of D (the *patterns*) where the support (card$\{g \in G \mid \forall d \in X, (g, d) \in I\}$) is greater than $minsupp$.

Then, formal concept analysis becomes a framework to mine frequent patterns. Indeed, it turns out that a formal concept captures the necessary information to represent a frequent pattern. The intent of a concept is a pattern and its extent is a set of objects that enjoy the attributes of the pattern. A frequent pattern determines a set of object whose cardinality is greater than $minsupp$. Mining frequent patterns thus amounts to navigate within the concept lattice to identify frequent patterns.

Not all frequent patterns belong to the concept lattice. However, the concept lattice contains enough information to derive the support of all frequent patterns. Indeed, a concept enjoys some form of closure: the intent is exactly the largest set of attributes shared by the objects of the extent; and the extent is exactly the largest set of objects which have the attributes of the intent.

5.4.4 Applying Pattern Structures to Chronicles

In this section, our objective is to apply pattern structures to chronicles. We start by defining a Galois connection between the poset of chronicles that conforms exactly with $\mathcal{E}_{\mathcal{D}}$ and the powerset of sequences in a dataset of sequences $\mathcal{D}$. Let us define two functions φ and ψ between $\widetilde{\mathcal{C}_{f\mathcal{E}_{\mathcal{D}}}}$ and the power set of examples as follows:

$$\varphi(\mathscr{C}) \stackrel{\text{def}}{=} \left\{ S \in Str(\mathcal{D}) \mid \mathscr{C} \preceq \Lambda_{\mathcal{E}_{\mathcal{D}}} \delta(S) \right\}$$

and

$$\psi(\mathcal{S}) \stackrel{\text{def}}{=} \underset{S \in \mathcal{S}}{\text{glb}} \, \Lambda_{\mathcal{E}_{\mathcal{D}}}(\delta(S)).$$

Proposition 5.5 (φ, ψ) *is a Galois connection between* $(\widetilde{\mathcal{C}_{f\mathcal{E}_{\mathcal{D}}}}, \preceq)$ *and* $(2^{Str(\mathcal{D})}, \supseteq)$.

In their approach, Ganter and Kuznetsov require the existence of a certain subsemilattice but they indicate a condition sufficient to ensure that: It amounts here to the finiteness of $\text{Im} \, \Lambda_{\mathcal{E}_{\mathcal{D}}} \circ \delta$, which happens to be an obvious fact (since $\mathcal{D}$ is finite, so is $\text{Im} \, \Lambda_{\mathcal{E}_{\mathcal{D}}} \circ \delta$).[2] This gives rise to the following proposition.

Proposition 5.6 *Let $\mathcal{D}$ be a dataset of sequences, then $(Str(\mathcal{D}), \widetilde{\mathcal{C}_{f\mathcal{E}_{\mathcal{D}}}}, \Lambda_{\mathcal{E}_{\mathcal{D}}} \circ \delta)$ is a pattern structure.*

Remark The definition of ψ does not match the definition of abstraction, *abs* (see Definition 5.5), introduced in the previous section. Indeed, an abstraction does not necessarily belong to the lattice of the chronicles that conform exactly with $\mathcal{E}_{\mathcal{D}}$. It need not either belong to a lattice $\widetilde{\mathcal{C}_{fP}}$.

---

[2] Sensu stricto, chronicles are even more general than labelled multigraphs because a chronicle may have infinitely many arcs between two vertices (e.g., the codomain of E would be $\mathbb{N} \cup \{\omega\}$ to allow for a countably infinite set of arcs between two vertices).

Example 5.32 (Formal Concepts) Example 5.28 illustrates the construction of the abstraction of the set of stripped sequences of our running example of a dataset $\mathcal{D}$ (see Example 5.25).

Figure 5.9 depicts the formal concepts extracted from this dataset $\mathcal{D}$. In this figure, the extent of a formal concept is the set of the sequences' identifiers. The dataset profile of $\mathcal{D}$ is $\mathcal{E}_{\mathcal{D}} = \{\{ABBCDD\}\}$. Then all chronicles share this multiset.

In pattern structures, the set of formal concepts is a lattice [47]. Figure 5.9 illustrates the structure of the lattice. A blue edge is a direct relationship between two formal concepts. The top concept corresponds to an extent that

(continued)

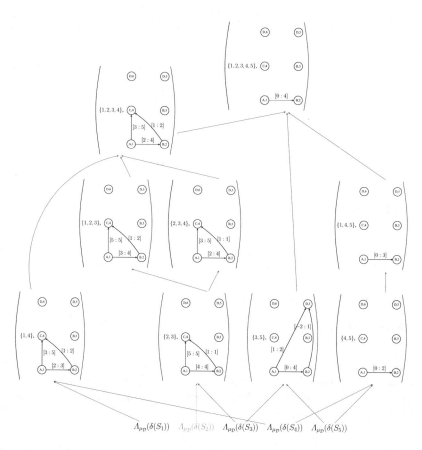

Fig. 5.9 Formal concepts extracted from the dataset of sequences of Example 5.25, p. 64. A formal concept is a pair in brackets. It contains a set of sequence indices and a chronicle. The blue arrows depict the sublattice of the powerset of stripped sequences. See Example 5.32 for more details

Example 5.32 (continued)

is the dataset. Then, this top concept identifies the similar chronicle as in the abstraction of the dataset (see Example 5.28). The bottom concept (not represented in the figure) is the concept with an empty extent. At the bottom of Fig. 5.9, we indicate the chronicle reducts of the sequences. These chronicles with their sequence identifier form a formal concept, except $\Lambda_{\mathcal{E}_{\mathcal{D}}}(\delta(S_2))$. The latter has the same chronicle as the formal concept whose extent is $\{2, 3\}$.

On the left part of the lattice, all chronicles have three temporal constraints between the same three events. They differ in the boundaries (the upper is the concept in the lattice, the larger are the intervals).

An interesting case is the formal concept with the extent $\{3, 5\}$: whatever sequence would be added to this extent, its generalisation into a chronicle occurs in all the sequences of the dataset. Indeed, this chronicle has a temporal constraint between $(A, 1)$ and $(B, 2)$ whose interval is the convex hull of all the delay between these events, but these two sequences have also $(D, 5)$ in common.

In the remaining of this section, we provide some results and remarks to characterize the chronicles identified by formal concepts in the pattern structure above.

The two following result relates the cardinal of the intent to the notion of support of chronicle. Then, it ensues that the chronicles in formal concepts are closed chronicles (see Definition 5.7).

Lemma 5.18 *Let $(\mathcal{C}, \mathcal{S})$ be a formal concept, then $supp_D(\mathcal{C}) = |\mathcal{S}|$*   □

Lemma 5.19 *The intent of a formal concept is a closed chronicle.*

Then, an alternative to the frequent chronicles mining algorithm is the use of formal concepts filtered by a constraint on the cardinal of their extent. The advantage of this approach is to extract only the set of closed chronicles which are less numerous than frequent chronicles but that contains the same information (all frequent chronicles and their support can be deduced from closed chronicles).

Remark In practice, we can imagine using the ⁻ operator to simplify the intents of the formal concepts in order to have simpler chronicle representation. Nonetheless, the simplified chronicle may occur in a larger set of sequences than the extent (see Example 5.33). This may be disturbing for a user familiar with formal concept analysis.

Example 5.33 (Intent Simplification Occurs in Larger Set than the Extent)
Let us consider the dataset consisting of three sequences: $\langle 1, \langle (A,4), (B,5) \rangle \rangle$, $\langle 2, \langle (A,3), (B,5) \rangle \rangle$ and $\langle 3, \langle (A,4), (A,5), (B,7) \rangle \rangle$.

Then, $C_1 = (\{1, 2\}, \mathscr{C}_1 = (\{\!\{A, A, B\}\!\}, \{(A, 1)[1 : 2](B, 3)\}))$ and $C_2 = (\{1, 2, 3\}, \mathscr{C}_2 = (\{\!\{A, A, B\}\!\}, \{(A, 1)[1 : 3](B, 3)\}))$ are two formal concepts.

If we look at the first concept, its simplification removes the second A, i.e., $\overline{\mathscr{C}_1} = (\{\!\{A, B\}\!\}, \{(A, 1)[1 : 2](B, 2)\})$. Then, $\overline{\mathscr{C}_1}$ occurs in sequences 1 and 2, but it also occurs in sequence 3. Indeed, if the events of $\overline{\mathscr{C}_1}$ maps the second A in the sequence and B, then the temporal constraint is satisfied. The concept with the extent $\{1, 2, 3\}$ is C_2, and as the temporal constraint considered involves the first A of the sequence, the intent of the concept is different from the one of C_1.

This example illustrates that care must be exercised when applying simplification to an extent.

Remark The reduction operation (Λ_μ) makes a choice among the possible embeddings of a chronicle in the sequence. Then, the reduction operation can be seen as a heuristics to guide the definition of temporal constraints when there are multiple possibilities. Clearly, this choice impacts the chronicles that are extracted as formal concepts.

Then, it is interesting to compare this choice with state of the art frequent chronicle mining that make alternative choices: HCDA [21] and Vu Duong [27]. These differences are illustrated in Example 5.34.

First, the approach we proposed focuses on the first occurrence of each event type. The reduction operation maps the events of a sequence to the events with the lowest indices in the multiset of the dataset. Then, temporal constraints are enforced to only involve events with the lowest indices (for each type of events). This can be observed in Fig. 5.6. As the extent of a formal concept is the glb of such a kind of chronicles, then it will admit temporal constraints only "where all the chronicles of the intent display temporal constraints", i.e., between events with the lowest indices.

In the approach of Vu Duong, the authors made a different choice. They prefer to extract chronicles with tight constraints. When they have to make a choice between several possible intervals, they select the interval $[l : u]$ such that $u - l$ is the smallest (tightest) and, if there is a tie, they select with the lowest bound value, i.e. $\max(|l|, |u|)$ is the smallest.

Finally, HCDA does not make any choice. The algorithm extracts the complete set of frequent chronicles using temporal constraints that occur in sequences.

Example 5.34 (Comparison with State of the Art Algorithms)
    Let us continue the previous example and compare sets of chronicles extracted by the three alternative approaches just discussed. We consider sets of chronicles whose support is strictly larger than 1.
    Our approach based on pattern structures extracts 3 chronicles (or formal concepts) with a support larger than 2. Example 5.33 gave two formal concepts, C_1 and C_2, obtained by our approach, the third formal concept with a support larger than 2 is $C_3 = (\{2, 3\}, \mathscr{C}_3 = (\{\{A, A, B\}\}, \{(A, 1)[2 : 3](B, 3)\}))$.
    The Vu Duong et al. [28] approach favours the chronicle with tight constraints. Thus, the approach extracts the chronicle $\mathscr{C}_1$ and $\mathscr{C}_3$ (with a support equals to 3 in this case). But it does not output the chronicle $\mathscr{C}_2$. This chronicle has the same multiset as $\mathscr{C}_1$ but the interval of the temporal constraint uses larger values ([2 : 3]) than the one of $\mathscr{C}_1$ ([1 : 2]). In this case, $\mathscr{C}_2$ is discarded.
    HCDA [21] extracts the three simplified chronicles but also the chronicle with temporal constraints ($\{\{A, B\}\}, \{(A, 1)[2 : 2](B, 2)\}$) that occurs in the second and third sequences.

Remark It is also interesting to compare our approach with the work of Sahuguède et al. [63] which proposes to represent chronicles as a polytope in high dimensional space. Both approaches seem very similar as the main idea is to embed the stripped sequences in a space that is sufficiently large to represent all the chronicles of the dataset. In the case of Sahuguède et al., this is a continuous space but, in our work, it is a discrete space.

5.5 Conclusion

In this section we use chronicle as a temporal model to abstract of a set of sequences. We first show that an event sequence can be represented by a chronicle. This embedding of the event sequences into the space of chronicles enables to exploit the structure of the space of chronicle to define some generalisation of a dataset of sequences.
    The core of the proposal of this chapter is the notion of *chronicle reduct*. A chronicle reduct enables to map the sequences of a dataset into a lattice of finite slim chronicles that exactly conform with the profile of the dataset. Then, a well-defined notion of abstraction of a dataset of sequences is derived from the lattice structure.
    More specifically, we proposed two different manners to abstract a dataset of sequences that corresponds to two different practical needs. Firstly, the *abstraction*

of a dataset of sequences is a single chronicle that occurs in each sequence of the dataset. The abstraction is derived from the unique chronicle that meets the reduct of each chronicle generated from the sequences. It can be seen as a summary of the dataset of sequences. Secondly, we proposed to abstract the dataset of sequences by a collection of chronicles. This task is also known as a pattern mining. Thanks to the lattice of chronicles we exhibited, the formal account of pattern structure applies. This lead us to propose a new approach for mining chronicles from a dataset of event sequences.

Our analysis of the chronicles concludes that chronicle reduct is a choice among the possible multiple embeddings of a chronicle in a stripped sequence. When there are multiple embeddings, the abstraction based on chronicle reduct considers the events of a given type with the lowest timestamps. The consequence for chronicle mining is that our framework based on pattern structure does not extract the same sets of chronicles than ones extracted by state-of-the-art chronicle mining algorithms. This may lead to conceive original algorithms with well-defined semantics.

Chapter 6
Conclusion

Abstract This chapter is the conclusion of the book. It simply gives a brief summary of its content and opens up toward new perspectives.

Keywords Chronicle · Discussion · Extensions

6.1 Summary

A lot of applications produce temporal sequences. Chronicles are of interest because they can be regarded as patterns to be found in such sequences. Intuitively, a chronicle consists of events with (possibly flexible) delays between them. For reasons ranging to ease of comparing two chronicles to practical issues in identifying occurrences of a chronicle in a temporal sequence, we do not define a chronicle simply as a multiset of events governed by delays but we introduce some further requirements. For one thing, we distinguish between the multiple instances of the same event type by labelling them with different indices. For another thing, we express delays as time intervals between two events ordered by their event type (that is, we disallow expressing a delay "the other way around" even in the case that this policy would cause mentioning the two events in reverse chronological order). In more detail, we enforce a chronicle to consist of (possibly duplicated) instances of event types together with time intervals specifying a delay between two of these instances, all of which is subject to respecting a fixed order upon event types.

After defining the notion of chronicle, we investigate the space of chronicles with the objective to identify a workable structure for using chronicles in the tasks aforementioned. More specifically, our initial objective was to identify a semi-lattice structure. For that, we define a pre-order $\preceq$ between chronicles and we investigate different subspaces of the space of chronicle to finally identify spaces of slim finite pairwise flush chronicles that enjoy a semi-lattice structure.

Regarding chronicles as patterns for temporal sequences, the notion of an occurrence of a chronicle in a temporal sequence arises quite naturally. It is then expected that chronicles could be compared through their occurrences in temporal sequences. Intuitively, a more general chronicle occurs in more temporal sequences. It is in practice unfeasible to check directly whether a chronicle occurs in more temporal sequences than another chronicle. Instead, the pre-order $\preceq$ between chronicles is of a syntactical nature (it only takes syntactical tests to check whether a chronicle is less than another) but is compatible with generality as just mentioned: For two chronicles $\mathscr{C}$ and $\mathscr{C}'$, if $\mathscr{C}' \preceq \mathscr{C}$ then every occurrence of $\mathscr{C}$ in a temporal sequence S is also an occurrence of $\mathscr{C}'$ in S. In other words, we define a workable sufficient condition for the relation "occurring in more temporal sequences". In addition, we investigated the notion of counting occurrences of a chronicle in a single sequence. The expected property of counting chronicles was again that a chronicle occurs more time in a sequence that the more specific ones. This requires to take a particular attention to multiple occurrences of a chronicle. For that, we proposed the notion of distinct occurrences which enjoys the desired property.

Regarding chronicles as an abstraction of temporal sequences, we first introduce how a sequence generates a chronicle such that abstracting temporal sequences leads to identify some of a least general generalisation of the temporal sequence. The semi-lattice structure of some chronicle subspaces defines naturally such kind of chronicle. The problem is that chronicles generated from temporal sequences belongs to different chronicle subspace. The address this problem, we propose the notion of chronicle reduct which enables to embed all chronicles generated from temporal sequences in the same semi-lattice. As such, we define the notion of abstraction of a dataset of temporal sequences and we also propose an original framework for chronicle mining. This framework relies on the notion of pattern structure, a generalisation of formal concept analysis.

6.2   Perspectives

All along this book, we identified potential limitations with our proposal that can be turned into an interesting open problem to address in future work. This section lists them. We organise them in several groups: the formal account of chronicles, restrictions due to the model of event sequence, possible increase in expressiveness, extension to multi-chronicle to address the limitation of the proposed approach for chronicle mining and finally investigating computational considerations.

We start by proposing some perspectives on the formal account of chronicle, i.e. mainly about the second chapter of this book.

About satisfiability of temporal constraint.   We have refrained from providing a formal notion of a temporal constraint being *satisfied* (by a pair of events, obviously). The reason is that the definition of a temporal constraint as being of the form $(e, o_e)[t^- : t^+](e', o_{e'})$ explicitly refers to an ordered multiset through the indices o_e and $o_{e'}$. Of course, this multiset is identified, because the temporal constraint under consideration depends on a chronicle, that has (by the definition) such a multiset. Nevertheless, the fact is: Events, as introduced by Definition 2.1, do not make sense of indices (whether in a multiset or elsewhere). It is to say that, e.g., $(A, 122.31)$ and $(B, 147.63)$ satisfy, for $t^- \leq 25.32 \leq t^+$, all temporal constraints $(A, i)[t^- : t^+](B, j)$ whatever i and j? Or is such a pair of events simply a potential case for all these temporal constraints to be satisfied? Or is a multiset embedding indispensable, in which case there cannot be any notion of a temporal constraint being satisfied by a pair events? All this seems rather intricate and perhaps the very notion is shaky. Accordingly, these questions underlie future work.

Simplification of slimming.   The need to check whether a chronicle is slim, and if not, to turn it into a slim equivalent, is bothersome. It is worth determining a choice of a system of representatives for the quotient set $C/ \approx$ as an alternative to finite slim chronicles. As a system of representatives is based on a choice function, various non-trivial issues are involved, starting with failure of associativity for intersection of chronicles. Even the existence of a such a system of representatives is not obvious. If there exist none, another direction would be to investigate the possibility of a weakening (fitted to pre-orders) for the notion of a Galois connection. Following such an approach, the idempotence of $f \circ g$ (and $g \circ f$) fails as only the weaker property that $g(x)$ is equivalent with $g(f(g(x)))$ (similarly, $f(x)$ equivalent with $f(g(f(x)))$) holds. However, chronicles we are considering form a finite set hence this weaker property ensures that the infinite series of equivalences starting with $g(x)$ and continuing with $g(f(g(x)))$ then $g(f(g(f(g(x)))))$ and so on ranges over a finite set hence at some point an identity is obtained. Such an identity can then be used for a solution to the representative issue just mentioned.

Conjunctive temporal constraints.   In a chronicle, the temporal constraints are interpreted conjunctively. For the chronicle to occur in a stripped sequence, some events in the stripped sequence must be as required by all the temporal constraints of the chronicle. It seems worth investigating the possibility of interpreting the temporal constraints in a chronicle disjunctively. Perhaps on a general level, or perhaps on a more restricted level, e.g., for temporal constraints between exactly the same pair of instances of event types in the chronicle: $(e_i, i)[l_1 : u_1](e_j, j), \ldots, (e_i, i)[l_k : u_k](e_j, j)$.

Toward the anonymity of instances of event types in a chronicle.   At   first sight, it seems natural that a multiset embedding is required to preserve the order between instances of the same event type (i.e., if the ith copy in $\mathcal{E}$ of A is mapped to the jth copy in $\mathcal{E}'$ of A, then the $i + 1$th copy in $\mathcal{E}$ of A cannot be mapped to the kth copy in $\mathcal{E}'$ of A for $k < j$). However, this turns out to be demanding to the point that some well-behaved chronicles get no occurrence at all. It would be interesting to look at circumstances that may allow lifting this requirement.

More generally, indexing instances of event types in a chronicle makes it rigid with respect to some aspects (please notice that indexing does have advantages regarding implementation issues). Here is an example. Imagine the case of a chronicle $\mathcal{C}$ that is supposed to express that two events of type A occur before an event of type B, with delays, say $[l_1 : u_1]$ and $[l_2 : u_2]$ (but there is no temporal requirement between the two events of type A other than induced by $[l_1 : u_1]$ and $[l_2 : u_2]$). Imagine that a second chronicle $\mathscr{C}'$ of the same kind (with delays $[l'_1 : u'_1]$ and $[l'_2 : u'_2]$) is to be considered, too. If $\mathscr{C}$ and $\mathscr{C}'$ were considered individually, it would not matter which instance of A (namely, $A, 1$ or $A, 2$) is required to occur before $B, 3$ within the delay $[l_i : u_i]$. Considering *both* $\mathscr{C}$ and $\mathscr{C}'$ is another story. Assume $[l_1 : u_1] \subseteq [l'_1 : u'_1]$ and $[l_2 : u_2] \subseteq [l'_2 : u'_2]$ while $[l_1 : u_1] \not\subseteq [l'_2 : u'_2]$ and $[l_2 : u_2] \not\subseteq [l'_1 : u'_1]$. Then, $\mathscr{C}' \preceq \mathscr{C}$ may or may not hold, depending on whether the temporal constraints are written $(A, i)[l_i : u_i](B, 3)$ in $\mathscr{C}$ and $\mathscr{C}'$ or the other way around, e.g., $(A, i)[l_i : u_i](B, 3)$ in $\mathscr{C}$ and $(A, 1)[l'_2 : u'_2](B, 3)$ and $(A, 2)[l'_1 : u'_1](B, 3)$ in $\mathscr{C}'$. This is to be contrasted with the intuition of the example, which has more to do with $\exists i, j \in \{1, 2\}, i \neq j$, $(A, i)[l_i : u_i](B, 3)$ and $(A, j)[l_j : u_j](B, 3)$ for $\mathscr{C}$ and similarly for $\mathscr{C}'$. In other words, "anonymity" of instances of event types is not possible: there is no way to capture *exactly* the case of Fig. 6.1 and this may be a topic for future work in relation with subgraph isomorphisms as mentioned in Sect. 2.4.

Infinite *vs* no temporal constraints.   In our account of chronicles, the case that, for some h and k, there is no temporal constraint between e_h and e_k differs from the case that $(e_h, h)[-\infty, +\infty](e_k, k)$ is the single temporal constraint between e_h and e_k. First, $(\mathcal{E}, \mathcal{T}) \preceq (\mathcal{E}, \mathcal{T} \cup \{(e_h, h)[-\infty, +\infty](e_k, k)\})$ but $(\mathcal{E}, \mathcal{T} \cup \{(e_h, h)[-\infty, +\infty](e_k, k)\}) \not\preceq (\mathcal{E}, \mathcal{T})$ whenever $\mathcal{T}$ contains no temporal constraint between e_h and e_k. If a more extensional account is preferred, then a change in definitions is needed.

Further, an extensional account of chronicles may allow having semi-definite interval. A semi-definite interval would specify a constraint only on one bound of the interval without defining the other bound. For some h and k, our formal account distinguishes having no constraint from having the constraint $(e_h, h)[-\infty, +\infty](e_k, k)$. Nonetheless, if it is possible to have the constraint $(e_h, h)[l, +\infty](e_k, k)$ to specify a lower bound on the temporal delay between e_h and e_k, it is not possible to specify that there is a constraint on the lower bound but no constraint on the upper bound. Again, this would need a change in our definitions.

We also identified some possible changes to investigate in the nature of the temporal data that a chronicle attempt to occur in:

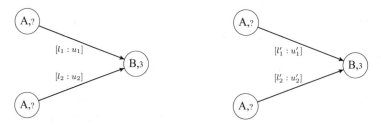

Fig. 6.1 Anonymity in chronicles

Handling multiplicity of event instances at the same time.   In   our   developments, a stripped sequence such as $\langle(A, 1.34)(A, 1.34)\rangle$ is not possible.
Nonetheless, it could be interesting to model the multiplicity of the event
instances. Here, A occurs twice at the same time. The problem with such
definition is that a chronicle can not match these two event instances because of
the strict order imposed on instances in the mapping image of event of the same
type. For instance, the chronicle $(\{\!\{A, A\}\!\}, \{(A, 1)[-1 : 1](A, 2)\})$ does not occur
in the sequence above. Then, it seems interesting to investigate the consequences
of relaxing the constraint on a strict order while keeping forbidding mapping of
two different events of the chronicle to the same instance in a sequence.

Another area for future work is streams.   They can be regarded as infinite
(stripped) sequences, going beyond our current account of temporal sequences.
Some delicate issues arise, such as the existence of infinitely many occurrences
of a chronicle.

Chapter 4 lays the groundwork by defining multiple occurrences of a chronicle
in a sequence. More specifically, by defining an anti-monotonic measure of the
amount of chronicles in a long sequence, it opens the way to adapt the formal
account of frequent chronicle mining we developed to the case of mining frequent
chronicles in a long sequence or a stream.

Presumably, the main area for improvements is about expressiveness. A first
improvement would consist in better characterizing the expressiveness of chronicles
in compare to logic formalism. Then, we suggest some potential practically useful
enhancement of the model that would be to formally investigate.

Comparing the expressiveness of chronicle wrt to temporal logics.   We briefly
compare chronicles with the model of Piel et al. [15]. As most of the classical
temporal logics (MTL, TPTL, etc.), the sequentiality of the instances of the
events in the formula is set. This is not the case with chronicles thanks to possible
negative delays. This raise a question: can any chronicle be expressed with a
formula of these logics? Our intuition is that chronicles can be expressed in TPTL
but not in MTL.

An avenue for extension concerns events with a duration.   An   avenue   for
extension concerns events with a duration. To deal with such events in an
approach disallowing duration, they are somehow split into two instantaneous
events, namely beginning and end. Of course, this is not optimal and a
development of chronicles to capture events with a duration is desirable.

Modelling absence in the event sequences.   In this book, a chronicle imposes
requirements such that in a certain time frame after an event of type X occurs, an
event of type Y is to occur. That is, a chronicle is limited to requirements about
events occurring. An improvement would be to permit requirements such that in
a certain time frame after the occurrence of an event of type X, no event of type
Y is to occur. Preliminary work in this direction are for instance [23, 41, 71].
A promising approach proposed by Dauxais et al. [23] defines absence as an
additional constraint between pairs of events. Then, more formal account have
to be developed to confirm the technical soundness of such approach. Indeed,

the problem of modelling absence has been studied in sequential patterns [9] and turns out to raise interesting questions about different semantics of absence.

Finally, working with multi-chronicles, i.e. a set of chronicles having the same multiset, may provide better generalisation of a dataset of chronicles. We have seen that abstraction of a chronicle based on chronicle reduct amounts to making a choice among the possible multiple embeddings of a chronicle in a sequence. A possibility to not impose the choice in the framework is to model the abstraction of a set of chronicle through a collection of chronicles, i.e. a multi-chronicle. Multi-chronicle may be seen similar to the notion of group of closed sequences introduced by Garriga [17] to summarize sequential data (without timestamps) with closed partial orders.

In addition to this possible future research directions about the formal properties of chronicles, a major aspect of temporal constraint networks is computational. Such a network is prone to computational change, towards an improved version, e.g., through constraint propagation. For example, the network $A[1, 2]B$, $B[1, 3]C$, $A[2, 9]C$ would be improved to $A[1, 2]B$, $B[1, 3]C$, $A[2, 5]C$, that is, changes are made to the network (of course, these changes do not alter the essence of the network). In our account, computation determines features of a chronicle, e.g., whether it is well-behaved, but computation entails *no change within the chronicle*. For example, the closure $\mathcal{T}^*$ is not supposed to replace $\mathcal{T}$ to give a better version of the original chronicle. A different point of view is that chronicles are really temporal constraint networks, in which case a notion of a better version would involve matters such as the requirement that there is a single temporal constraint between two nodes of the chronicle. This would be an avenue for further work as a number of properties depend on such an assumption.

Furthermore, no analysis is provided as to the cost of checking occurrences of a chronicle in sequences and other tasks. It would be useful to determine the computational complexity of these tasks. This analysis would be interesting to develop efficient tools that would enable a broad audience to use chronicles.

Appendix A
Proofs

A.1 Proofs for Sect. 2.1.2

Lemma 2.2 Let $\mathscr{C} = (\mathcal{E}, \mathcal{T})$ be a chronicle where $\mathcal{E} = \{\!\{e_1, \ldots, e_n\}\!\}$. Then, for all $(e_i, o)[t^- : t^+](e_j, p)$ in $\mathcal{T}$, it is the case that $e_o = e_i$ and $e_p = e_j$.

Proof of Lemma 2.2 By Definition 2.2, $e_{o_e} = e$ and $e_{o_{e'}} = e'$ which can be applied here for o_e being o and $o_{e'}$ being p, resulting in $e_o = e_i$ and $e_p = e_j$. □

Lemma 2.3 $\mathcal{E}' \Subset \mathcal{E}$ iff for all $e \in \mathbb{E}$, $\mu_{\mathcal{E}'}(e) \leq \mu_{\mathcal{E}}(e)$.

Proof of Lemma 2.3 Let us write $\mathcal{E}$ as $\{\!\{e_1, \ldots, e_m\}\!\}$ and write $\mathcal{E}'$ as $\{\!\{e_1', \ldots, e_{m'}'\}\!\}$.

($\Rightarrow$) Assume $\mathcal{E}' \Subset \mathcal{E}$. Due to Definition 2.6, there exists an increasing function $\theta : [m'] \to [m]$ such that $e_i' = e_{\theta(i)}$. The case $\mu_{\mathcal{E}'}(e) = 0$ implies $\mu_{\mathcal{E}'}(e) \leq \mu_{\mathcal{E}}(e)$. For $k \geq 1$, if $\mu_{\mathcal{E}'}(e) = k$ then $e = e_i' = e_{i+1}' = \ldots = e_{i+k-1}'$ for some i (Definition 2.2 imposes $e_1' \leq_{\mathbb{E}} \cdots \leq_{\mathbb{E}} e_{m'}'$). Accordingly, $e_{\theta(i)} = \ldots = e_{\theta(i+k-1)}$. Since θ is increasing, θ is injective and it henceforth follows that $\mu_{\mathcal{E}}(e) \geq k$.

($\Leftarrow$) Assume $\mu_{\mathcal{E}'}(e) \leq \mu_{\mathcal{E}}(e)$ for all $e \in \mathbb{E}$. Define, for $i \in [m']$,

$$
\theta(i) = \begin{cases} 1 + \sum_{e_j <_{\mathbb{E}} e_i'} \mu_{\mathcal{E}}(e_j) & \text{if } i = 1 \text{ or } e_{i-1}' \neq e_i' \\ 1 + \theta(i-1) & \text{else (i.e., } e_{i-1}' = e_i') \end{cases}
$$

This is well-defined because the argument in the second case is decreasing and the first case is a non-recursive value. Due to $\mu_{\mathcal{E}'}(e) \leq \mu_{\mathcal{E}}(e)$, $\mu_{\mathcal{E}}(e_i') \geq 1$, i.e., e_i' occurs in $\{\!\{e_1, \ldots, e_m\}\!\}$. Since $e_1 \leq_{\mathbb{E}} \cdots \leq_{\mathbb{E}} e_m$, the length of the maximal initial proper segment of $\{\!\{e_1, \ldots, e_m\}\!\}$ that does not include e_i' is $l = \sum_{e_j <_{\mathbb{E}} e_i'} \mu_{\mathcal{E}}(e_j)$. Thus, e_{l+1}, the next item after the segment is e_i' hence $e_i' = e_{1+l}$. If $i = 1$ or $e_{i-1}' \neq e_i'$ then $\theta(i) = 1 + \sum_{e_j <_{\mathbb{E}} e_i'} \mu_{\mathcal{E}}(e_j) = 1 + l$, therefore, $e_i' = e_{\theta(i)}$. Else, there exists $k \leq i - 1$ such that $e_k' = e_{k+1}' = \ldots = e_{i-1}' = e_i'$ and either $k = 1$ or $e_{k-1}' \neq e_k'$. That is, e_i' occurs at least $i - k + 1$ times in $\mathcal{E}'$. Then, $\theta(i) = (i-k) + \theta(k)$.

© The Author(s), under exclusive license to Springer Nature Switzerland AG 2023
T. Guyet, P. Besnard, *Chronicles: Formalization of a Temporal Model*,
SpringerBriefs in Computer Science, https://doi.org/10.1007/978-3-031-33693-5

However, $\theta(k) = 1 + \sum_{e_j <_{\mathbb{E}} e'_k} \mu_{\mathcal{E}}(e_j) = 1 + \sum_{e_j <_{\mathbb{E}} e'_i} \mu_{\mathcal{E}}(e_j) = 1 + l$. Substituting, $\theta(i) = (i - k) + 1 + l$. From $e'_k = e'_{k+1} = \ldots = e'_{i-1} = e'_i$ and $e'_i = e_{l+1}$ and $\mu_{\mathcal{E}'}(e) \leq \mu_{\mathcal{E}}(e)$, it follows that $e_{l+1} = e_{l+2} = \ldots = e_{i-k+l-1} = e_{i-k+l+1}$ (remember: $l = 0$ or $e_l <_{\mathbb{E}} e'_i$). Therefore, $e'_i = e_{l+1}$ gives $e'_i = e_{i-k+l+1} = e_{\theta(i)}$.

There remains to show that θ is increasing. The case that $e'_{i+1} = e'_i$ is trivial: $\theta(i + 1) = 1 + \theta(i)$. The other case is $e'_{i+1} \neq e'_i$. Hence, $\theta(i + 1) = 1 + \sum_{e_j <_{\mathbb{E}} e'_{i+1}} \mu_{\mathcal{E}}(e_j)$. From $e'_{i+1} \neq e'_i$ and $e'_i \leq_{\mathbb{E}} e'_{i+1}$, it ensues that $\theta(i + 1) = 1 + \mu_{\mathcal{E}}(e'_i) + \sum_{e_j <_{\mathbb{E}} e'_i} \mu_{\mathcal{E}}(e_j)$. Assume further $e'_{i-1} = e'_i$. There exists $k \leq i - 1$ such that $e'_k = e'_{k+1} = \ldots = e'_{i-1} = e'_i$ and either $k = 1$ or $e'_{k-1} \neq e'_k$. That is, e'_i occurs exactly $i - k + 1$ times in $\mathcal{E}'$. Thus, $\theta(i) = (i - k) + \theta(k)$. However, $\theta(k) = 1 + \sum_{e_j <_{\mathbb{E}} e'_k} \mu_{\mathcal{E}}(e_j) = 1 + \sum_{e_j <_{\mathbb{E}} e'_i} \mu_{\mathcal{E}}(e_j)$. Then, $\theta(i+1) = 1 + \mu_{\mathcal{E}}(e'_i) + \sum_{e_j <_{\mathbb{E}} e'_i} \mu_{\mathcal{E}}(e_j)$ gives $\theta(i+1) = \mu_{\mathcal{E}}(e'_i) + \theta(k) \geq \mu_{\mathcal{E}'}(e'_i) + \theta(k)$ by the assumption $\mu_{\mathcal{E}'}(e) \leq \mu_{\mathcal{E}}(e)$. Hence $\theta(i + 1) \geq (i - k + 1) + \theta(k) = 1 + \theta(i)$. Otherwise, assume further $e'_{i-1} \neq e'_i$. Consequently, $\theta(i) = 1 + \sum_{e_j <_{\mathbb{E}} e'_i} \mu_{\mathcal{E}}(e_j)$. In view of $\theta(i + 1) = 1 + \mu_{\mathcal{E}}(e'_i) + \sum_{e_j <_{\mathbb{E}} e'_i} \mu_{\mathcal{E}}(e_j)$, it ensues that $\theta(i + 1) = \mu_{\mathcal{E}}(e'_i) + \theta(i)$. Now, $\mu_{\mathcal{E}}(e'_i) \geq 1$ in view of the obvious fact $\mu_{\mathcal{E}'}(e'_i) \geq 1$ and the assumption $\mu_{\mathcal{E}'}(e) \leq \mu_{\mathcal{E}}(e)$. Thus, $\theta(i + 1) \geq 1 + \theta(i)$. $\square$

Lemma 2.4 $\mathcal{E} \cap \mathcal{E}'$ is the glb of $\mathcal{E}$ and $\mathcal{E}'$ wrt $\subseteq$.

Proof of Lemma 2.4 Consider $\mathcal{E}''$ such that $\mathcal{E}'' \subseteq \mathcal{E}$ and $\mathcal{E}'' \subseteq \mathcal{E}'$. In view of Lemma 2.3, both $\mu_{\mathcal{E}''}(e) \leq \mu_{\mathcal{E}}(e)$ and $\mu_{\mathcal{E}''}(e) \leq \mu_{\mathcal{E}'}(e)$ for all $e \in \mathbb{E}$. Therefore, $\mu_{\mathcal{E}''}(e) \leq \min(\mu(e), \mu'(e)) = \mu_{\mathcal{E} \cap \mathcal{E}'}(e)$, that is, $\mathcal{E}'' \subseteq \mathcal{E} \cap \mathcal{E}'$ due to Lemma 2.3 again. The other condition, $\mathcal{E} \cap \mathcal{E}' \subseteq \mathcal{E}$ and $\mathcal{E} \cap \mathcal{E}' \subseteq \mathcal{E}'$, is obvious. $\square$

Lemma 2.5 $\preceq$ is a preorder.

Proof of Lemma 2.5 Reflexivity is trivial. As regards transitivity, assume $\mathcal{C}' \preceq \mathcal{C}$ and $\mathcal{C}'' \preceq \mathcal{C}'$. From the latter, there exists θ' such that for all $(e''_i, o)[l'' : u''](e''_j, p)$ in $\mathcal{T}''$ there exists a temporal constraint $(e'_{\theta'(i)}, \theta'(o))[l' : u'](e'_{\theta'(j)}, \theta'(p))$ in $\mathcal{T}'$ that satisfies $(e'_{\theta'(i)}, \theta'(o))[l' : u'](e'_{\theta'(j)}, \theta'(p)) \trianglelefteq (e''_i, o)[l'' : u''](e''_j, p)$. There also exists θ such that for all $(e'_i, o)[l' : u'](e'_j, p)$ in $\mathcal{T}'$ there exists, in $\mathcal{T}$, a temporal constraint $(e_{\theta(i)}, \theta(o))[l : u](e_{\theta(j)}, \theta(p))$ s.t. $(e_{\theta(i)}, \theta(o))[l : u](e_{\theta(j)}, \theta(p)) \trianglelefteq (e'_i, o)[l' : u'](e'_j, p)$. Combining, there exists $(e_{\theta(\theta'(i))}, \theta(\theta'(o)))[l : u](e_{\theta(\theta'(j))}, \theta(\theta'(p)))$ in $\mathcal{T}$ that satisfies $(e_{\theta(\theta'(i))}, \theta(\theta'(o)))[l : u](e_{\theta(\theta'(j))}, \theta(\theta'(p))) \trianglelefteq (e'_{\theta'(i)}, \theta'(o))[l' : u'](e'_{\theta'(j)}, \theta'(p))$. Then, there remains to show that $\theta \circ \theta'$ is a multiset embedding and that $\trianglelefteq$ is transitive. The former holds in view of the notion of a multiset in Definition 2.2 and the latter is obvious from the notion of a temporal constraint in Definition 2.2. $\square$

Lemma 2.6 Let $\mathcal{C} = (\mathcal{E}, \mathcal{T})$ and $\mathcal{C}'' = (\mathcal{E}'', \mathcal{T}'')$ be chronicles such that $\mathcal{C}'' \preceq \mathcal{C}$. If $\mathcal{E}'' \subseteq \mathcal{E}' \subseteq \mathcal{E}$ then $(\mathcal{E}', \mathcal{T}'') \preceq \mathcal{C}$.

Proof of Lemma 2.6 Let us write $\mathcal{E} = \{\!\{e_1, \ldots, e_n\}\!\}$ and $\mathcal{E}'' = \{\!\{e''_1, \ldots, e''_m\}\!\}$. Also, $\mathcal{E}' = \{\!\{e'_1, \ldots, e'_s\}\!\}$. Considering a multiset embedding $\theta : [m] \to [n]$ for $\mathcal{E}'' \subseteq \mathcal{E}$, define $\vartheta : \{0, 1, \ldots, s\} \to [n]$ (although we only consider its restriction $\vartheta : [s] \to$

$[n]$ when examining $\preceq$ below) as follows:

$$\vartheta(h) = \begin{cases} \sum \{\mu_{\mathcal{E}}(e) \mid e <_{\mathbb{E}} e'_1\} & \text{if } h = 0 \\ \theta(k) & \text{if } e'_h = e''_k \text{ and, for } D_k \\ & \qquad \stackrel{\text{def}}{=} \{d \in [m-1] \mid e''_k = e''_{k-d}\}, \\ & \qquad h = 1 + \sup D_k + \sum \{\mu_{\mathcal{E}'}(e) \mid e <_{\mathbb{E}} e''_k\} \\ 1 + \sum \{\mu_{\mathcal{E}}(e) \mid e <_{\mathbb{E}} e'_h\} & \text{if } e'_h \neq e'_{h-1} \text{ and } e'_h \neq e''_k \text{ for all } k \\ \vartheta(h-1) + 1 & \text{else} \end{cases}$$

ϑ is well-defined because k in clause 2 is unique whenever it exists: should $l > k$ both satisfy the condition of clause 2 for the same h, then $e''_l = e'_h = e''_k$ while $\sup D_k < \sup D_l$ in view of $e''_1 \leq_{\mathbb{E}} \cdots \leq_{\mathbb{E}} e''_m$ and it follows that $h = 1 + \sup D_k + \sum \{\mu_{\mathcal{E}'}(e) \mid e <_{\mathbb{E}} e'_h\} < 1 + \sup D_l + \sum \{\mu_{\mathcal{E}'}(e) \mid e <_{\mathbb{E}} e'_h\} = h$, a contradiction.

The statement in the lemma assumes that $\mathscr{C}''$ is a lower bound of $\mathscr{C}$. There thus exists some multiset embedding for $\mathcal{E}'' \in \mathcal{E}$, that is, a strictly increasing $\theta : [m] \to [n]$ such that $e_{\theta(i)} = e''_i$ for $i = 1, \ldots, m$ and satisfying the condition in Definition 2.8. Instantiating θ in the above definition of ϑ by this multiset embedding, please observe that $e'_h = e_{\vartheta(h)}$ by construction (indeed, as regards clause 3, $1 + \sum \{\mu_{\mathcal{E}}(e) \mid e <_{\mathbb{E}} e'_h\}$ is the least index h_0 such that $e'_{h_0} = e'_h$ hence $e_{\vartheta(h)} = e'_h$).

Moreover, θ is (due to $\mathscr{C}'' \preceq \mathscr{C}$) such that for all $(e''_i, i)[l'' : u''](e''_j, j)$ in $\mathcal{T}''$, there exists $(e_{\theta(i)}, \theta(i))[l : u](e_{\theta(j)}, \theta(j))$ (denoted τ) in $\mathcal{T}$ satisfying $e_{\theta(i)} = e''_i$ and $e_{\theta(j)} = e''_j$ and $[l : u] \subseteq [l'' : u'']$. Our next step is to prove that τ can be written $(e_{\vartheta(h)}, \vartheta(h))[l : u](e_{\vartheta(w)}, \vartheta(w))$ for some h and w. Consider now $e''_i \in \mathcal{E}''$. There are two cases. Assume first $e''_{i-1} = e''_i$. Since $e''_1 \leq_{\mathbb{E}} \cdots \leq_{\mathbb{E}} e''_m$, the greatest d such that $e''_{i-d} = e''_i$ exists. Accordingly, $1 + d + \sum \{\mu_{\mathcal{E}'}(e) \mid e <_{\mathbb{E}} e''_i\}$ is an appropriate h to satisfy the condition for clause 2 in the definition of ϑ. Second, assume instead $e''_{i-1} \neq e''_i$ (possibly from the degenerate case $i = 1$). D_k is thus empty and $\sup D_k = 0$. Then, $d = 0$, and $1 + d + \sum \{\mu_{\mathcal{E}'}(e) \mid e <_{\mathbb{E}} e''_i\}$ is an appropriate h to satisfy the condition for clause 2 in the definition of ϑ. In either case, $\vartheta(1 + d + \sum \{\mu_{\mathcal{E}'}(e) \mid e <_{\mathbb{E}} e''_i\}) = \theta(i)$. Then, for $h = 1 + \sum \{\mu_{\mathcal{E}'}(e) \mid e <_{\mathbb{E}} e''_i\}$, τ can be written $(e_{\vartheta(h)}, \vartheta(h))[l : u](e_{\theta(j)}, \theta(j))$. As $e_{\theta(i)} = e''_i$, we have $e_{\vartheta(h)} = e''_i$. The rightmost part of $(e_{\theta(i)}, \theta(i))[l : u](e_{\theta(j)}, \theta(j))$ can be handled similarly so that τ can be written $(e_{\vartheta(h)}, \vartheta(h))[l : u](e_{\vartheta(w)}, \vartheta(w))$ while $e_{\vartheta(h)} = e''_i$ and $e_{\vartheta(w)} = e''_j$. In view of $[l : u] \subseteq [l'' : u'']$, it thus ensues that $(e_{\vartheta(h)}, \vartheta(h))[l : u](e_{\vartheta(w)}, \vartheta(w)) \trianglelefteq (e''_i, i)[l'' : u''](e''_j, j)$. Moreover, it follows that $(e_{\vartheta(h)}, \vartheta(h))[l : u](e_{\vartheta(w)}, \vartheta(w)) \trianglelefteq (e'_h, h)[l'' : u''](e'_w, w)$ due to $e'_h = e_{\vartheta(h)} = e''_i$ and $e'_w = e_{\vartheta(w)} = e''_j$.

There remains to show that ϑ is strictly increasing. Consider $\vartheta(h+1)$ where $h \geq 1$. There are various cases.

- If $e'_{h+1} \neq e'_h$ then $e'_h <_\mathbb{E} e'_{h+1}$ because $e'_1 \leq_\mathbb{E} \cdots \leq_\mathbb{E} e'_s$. However, $e'_{h+1} = e_{\vartheta(h+1)}$ and $e'_h = e_{\vartheta(h)}$. Then, $e_{\vartheta(h)} <_\mathbb{E} e_{\vartheta(h+1)}$. In view of $e_1 \leq_\mathbb{E} \cdots \leq_\mathbb{E} e_n$, it thus follows that $\vartheta(h) < \vartheta(h+1)$.
- Otherwise, $e'_{h+1} = e'_h$. In all subcases under this assumption, clause 3 fails to apply for $h + 1$ $(\star)$.

  - Assume further $\mu_{\mathcal{E}''}(e'_h) = 0$. As a consequence, $e'_h \neq e''_k$ for all k. Equivalently, $e'_{h+1} \neq e''_k$ for all k and clause 2 fails to apply for $h + 1$. By $(\star)$, there only remains clause 4 to apply for $h + 1$, and $\vartheta(h) < \vartheta(h+1)$ results.
  - Assume $\mu_{\mathcal{E}''}(e'_h) \neq 0$ instead. Since $e''_1 \leq_\mathbb{E} \cdots \leq_\mathbb{E} e''_m$, the least k_0 exists such that $e''_{k_0} = e''_{k_0+1} = \cdots = e'_h$. Obviously, $D_{k_0} = \emptyset$. Thus,

    $1 + \sup D_{k_0} + \sum \left\{ \mu_{\mathcal{E}'}(e) \mid e <_\mathbb{E} e''_{k_0} \right\} = 1 + \sum \left\{ \mu_{\mathcal{E}'}(e) \mid e <_\mathbb{E} e''_{k_0} \right\} = 1 + \sum \left\{ \mu_{\mathcal{E}'}(e) \mid e <_\mathbb{E} e'_h \right\}$ is the least index h_0 such that $e'_{h_0} = e''_{k_0}$ (as $e'_1 \leq_\mathbb{E} \cdots \leq_\mathbb{E} e'_s$). Then, $h_0 \leq h < h + 1$.

    · Should clause 2 fail to apply for $h + 1$, $(\star)$ would imply that there only remains clause 4 to apply for $h + 1$, and $\vartheta(h) < \vartheta(h+1)$ results.
    · The last case is that clause 2 applies for $h + 1$ through some k_{h+1}. Thus,

    $$e'_{h+1} = e''_{k_{h+1}}$$

    and

    $$h + 1 = 1 + \sup D_{k_{h+1}} + \sum \left\{ \mu_{\mathcal{E}'}(e) \mid e <_\mathbb{E} e''_{k_{h+1}} \right\}.$$

    Using the equalities established just above, namely $e''_{k_0} = e''_{k_0+1} = \cdots = e'_h$ and $e'_{h_0} = e''_{k_0}$ together with the lasting assumption $e'_{h+1} = e'_h$, it follows that

    $$e''_{k_{h+1}} = e'_{h+1} = e'_h = e''_{k_0} = e'_{h_0}$$

    and

    $$h + 1 = 1 + \sup D_{k_{h+1}} + \sum \left\{ \mu_{\mathcal{E}'}(e) \mid e <_\mathbb{E} e''_{k_0} \right\} \tag{0}$$

    In view of $e'_{h_0} = e''_{k_0}$ and the fact that h_0 is the least index such that $e'_{h_0} = e''_{k_0}$, $\sum \left\{ \mu_{\mathcal{E}'}(e) \mid e <_\mathbb{E} e''_{k_0} \right\} = h_0 - 1$. Substituting in (0), $\sup D_{k_{h+1}} = h - h_0 + 1$. By the definition of $D_{k_{h+1}}$ (it becomes $D_{k_{h+1}} = \{d \in [m-1] \mid e''_{k_0} = e''_{k_{h+1}-d}\}$ due to the above equalities) and by the fact that k_0 is the least index such that $e''_{k_0} = e'_h$ (also, keep in mind that $e''_{k_0} = e''_{k_0+1} = \cdots = e''_{k_{h+1}}$), it holds that

$$k_{h+1} = k_0 + h - h_0 + 1.$$

Let us show that $k_0 + h - h_0$ is an appropriate k for clause 2 to apply for h. First, the facts that $e''_{k_0+h-h_0+1} = e''_{k_0}$ and $k_0 \leq k_0+h-h_0 < k_0+h-h_0+1$ and $e''_{k_0} \leq_{\mathbb{E}} e''_{k_0+1} \leq_{\mathbb{E}} \cdots$ together imply $e''_{k_0+h-h_0} = e''_{k_0}$. Since $e'_h = e''_{k_0}$,

$$e'_h = e''_{k_0+h-h_0} = e''_k$$

which ensures the first part of clause 2 for h. As to the second part, consider

$$1 + \sup D_k + \sum \left\{ \mu_{\mathcal{E}'}(e) \mid e <_{\mathbb{E}} e''_k \right\}.$$

As $e''_k = e'_h$ was just shown and $e'_h = e''_{k_0}$, this becomes

$$1 + \sup D_k + \sum \left\{ \mu_{\mathcal{E}'}(e) \mid e <_{\mathbb{E}} e''_{k_0} \right\}$$

and for the same reasons as above, $\sum \left\{ \mu_{\mathcal{E}'}(e) \mid e <_{\mathbb{E}} e''_{k_0} \right\} = h_0 - 1$. We have shown $D_{k_{h+1}} = \{d \in [m-1] \mid e''_{k_0} = e''_{k_{h+1}-d}\}$. However, $k_{h+1} = k + 1$. Then, $D_k = D_{k_{h+1}} \setminus \{h - h_0 + 1\}$ (due to $D_k = \{d \in [m-1] \mid e''_{k_0} = e''_{k-d}\}$ together with $\sup D_{k_{h+1}} = h - h_0 + 1$). Therefore, $\sup D_k = h - h_0$. Summing it up,

$$1 + \sup D_k + \sum \left\{ \mu_{\mathcal{E}'}(e) \mid e <_{\mathbb{E}} e''_k \right\}$$
$$= 1 + \sup D_k + \sum \left\{ \mu_{\mathcal{E}'}(e) \mid e <_{\mathbb{E}} e''_{k_0} \right\}$$
$$= 1 + \sup D_k + h_0 - 1$$
$$= 1 + h - h_0 + h_0 - 1$$
$$= h$$

Thus, clause 2 applies for h via $k = k_0+h-h_0$ hence $\vartheta(h) = \theta(k_0+h-h_0)$ whereas clause 2 applies for $h+1$ via $k_{h+1} = k_0 + h - h_0 + 1$ so that $\vartheta(h+1) = \theta(k_0 + h - h_0 + 1)$. Since θ is strictly increasing, $\vartheta(h) < \vartheta(h+1)$.

This concludes the proof. $\qquad\square$

Lemma 2.8 If $(\mathcal{E}, \mathcal{T}') \preceq (\mathcal{E}, \mathcal{T})$ then $(\mathcal{E}, \mathcal{T} \cup \mathcal{T}') \preceq (\mathcal{E}, \mathcal{T})$.

Proof of Lemma 2.8 Let us write $\mathcal{E}$ as $\{\!\{e_1, \ldots, e_n\}\!\}$. Due to $(\mathcal{E}, \mathcal{T}') \preceq (\mathcal{E}, \mathcal{T})$, Definition 2.8 expresses that there exists a multiset embedding θ which ensures (1) $\mathcal{E} \subseteq \mathcal{E}$ and (2) for each temporal constraint $(e_i, o')[l' : u'](e_j, p')$ in $\mathcal{T}'$, there exists a temporal constraint $(e_{\theta(i)}, \theta(o'))[l : u](e_{\theta(j)}, \theta(p'))$ in $\mathcal{T}$ such that $(e_{\theta(i)}, \theta(o'))[l : u](e_{\theta(j)}, \theta(p')) \unlhd (e_i, o')[l' : u'](e_j, p')$. However, θ must be from $[n]$ to $[n]$ and must be strictly increasing. There is only one solution: θ is identity over $[n]$, i.e., $e_h = e_{\theta(h)}$ for all $h \in [n]$. Substituting in (2) then gives: for each temporal constraint $(e_i, o')[l' : u'](e_j, p')$ in $\mathcal{T}'$, there exists a temporal constraint $(e_i, o')[l : u](e_j, p')$ in $\mathcal{T}$ such that $(e_i, o')[l : u](e_j, p') \unlhd (e_i, o')[l' : u'](e_j, p')$.

We must show $(\mathcal{E}, \mathcal{T} \cup \mathcal{T}') \preceq (\mathcal{E}, \mathcal{T})$. Consider $(e_i, o')[l' : u'](e_j, p')$ a temporal constraint in $\mathcal{T}'$. As just shown, there exists a temporal constraint $(e_i, o')[l : u](e_j, p')$ in $\mathcal{T}$ such that $(e_i, o')[l : u](e_j, p') \unlhd (e_i, o')[l' : u'](e_j, p')$. For $(e_i, o)[l : u](e_j, p)$ a temporal constraint in $\mathcal{T}$, it is clear that $(e_i, o)[l : u](e_j, p) \unlhd (e_i, o)[l : u](e_j, p)$. Hence, for θ being identity over $[n]$, we have shown that each temporal constraint in $\mathcal{T} \cup \mathcal{T}'$ is $\unlhd$-dominating as required in Definition 2.8. Moreover, it is obvious that θ as identity over $[n]$ is a multiset embedding for $\mathcal{E} \Subset \mathcal{E}$.

$\square$

A.2   Proofs for Sect. 3.2

Proposition 3.1 $(\widetilde{\mathcal{C}}, \preceq)$ is a poset.

Proof of Proposition 3.1 By Lemma 2.5, there only remains to show that $\preceq$ over $\widetilde{\mathcal{C}}$ is antisymmetric. Consider two slim chronicles $\mathscr{C} = (\mathcal{E}, \mathcal{T})$ and $\mathscr{C}' = (\mathcal{E}', \mathcal{T}')$, writing $\mathcal{E}$ as $\{\!\{e_1, \ldots, e_m\}\!\}$ and $\mathcal{E}'$ as $\{\!\{e'_1, \ldots, e'_{m'}\}\!\}$. Assume $\mathscr{C}' \preceq \mathscr{C}$ and $\mathscr{C} \preceq \mathscr{C}'$. Clearly, $\mathcal{E}' \Subset \mathcal{E}$ and $\mathcal{E} \Subset \mathcal{E}'$. Due to Lemma 2.3, $\mathcal{E}' \Subset \mathcal{E}$ means $\mu_{\mathcal{E}'}(e) \le \mu_{\mathcal{E}}(e)$ for all $e \in \mathbb{E}$ while $\mathcal{E} \Subset \mathcal{E}'$ means $\mu_{\mathcal{E}}(e) \le \mu_{\mathcal{E}'}(e)$ for all $e \in \mathbb{E}$. Trivially, $\mathcal{E} = \mathcal{E}'$ ensues. Then, $m = m'$. This implies that a multiset embedding θ for $\mathcal{E}' \Subset \mathcal{E}$ must be from $[m]$ to $[m]$ and must be strictly increasing, thus θ must be identity over $[m]$.

Let us exploit the assumption $\mathscr{C}' \preceq \mathscr{C}$. Due to Definition 2.8, $\mathcal{E}' \Subset \mathcal{E}$ by some multiset embedding θ such that for every temporal constraint $(e'_i, o')[l' : u'](e'_j, p')$ in $\mathcal{T}'$ there exists a temporal constraint $(e_{\theta(i)}, \theta(o'))[l : u](e_{\theta(j)}, \theta(p'))$ in $\mathcal{T}$ that satisfies $(e_{\theta(i)}, \theta(o'))[l : u](e_{\theta(j)}, \theta(p')) \unlhd (e'_i, o')[l' : u'](e'_j, p')$ where $e'_h = e_{\theta(h)}$ for $h = 1, \ldots, m'$. So, for all $(e'_i, o')[l' : u'](e'_j, p')$ in $\mathcal{T}'$, there exists $(e'_i, o')[l : u](e'_j, p')$ in $\mathcal{T}$ such that $[l : u] \subseteq [l' : u']$.

Similarly, a multiset embedding θ' for $\mathcal{E} \Subset \mathcal{E}'$ must be identity over $[m']$ (i.e., $[m]$). Symmetrically to the case above, $\mathscr{C} \preceq \mathscr{C}'$ implies that for every $(e_i, o)[l : u](e_j, p)$ in $\mathcal{T}$, there exists $(e_i, o)[l' : u'](e_j, p)$ in $\mathcal{T}'$ such that $[l' : u'] \subseteq [l : u]$.

Combining the two cases results in the property: If $(e'_i, o')[l' : u'](e'_j, p') \in \mathcal{T}'$, there exists $(e'_i, o')[l : u](e'_j, p')$ in $\mathcal{T}$ such that $[l : u] \subseteq [l' : u']$, while, in turn, for this $(e'_i, o')[l : u](e'_j, p')$ in $\mathcal{T}$, there exists $(e'_i, o')[l'' : u''](e'_j, p')$ in $\mathcal{T}'$ such that $[l'' : u''] \subseteq [l : u]$. Clearly, $[l'' : u''] \subseteq [l' : u']$ ensues. The assumption that $\mathscr{C}'$ is a slim chronicle then implies that $(e'_i, o')[l' : u'](e'_j, p')$ and $(e'_i, o')[l'' : u''](e'_j, p')$ are the same temporal constraint. Thus, $l' = l''$ and $u' = u''$. Then, $[l : u] \subseteq [l' : u']$ and $[l' : u'] \subseteq [l : u]$ henceforth $l = l'$ and $u = u'$. That is, $(e'_i, o')[l : u](e'_j, p')$

is also the same temporal constraint as $(e_i', o')[l' : u'](e_j', p')$. Stated otherwise, if $(e_i', o')[l' : u'](e_j', p')$ is in $\mathcal{T}'$ then it is in $\mathcal{T}$. This shows $\mathcal{T}' \subseteq \mathcal{T}$.

Of course, $\mathcal{T} \subseteq \mathcal{T}'$ can be shown similarly. Overall, $\mathcal{T} = \mathcal{T}'$. Since $\mathcal{E} = \mathcal{E}'$ has already been shown, $\mathscr{C} = \mathscr{C}'$. □

Lemma 3.3 $\widetilde{\mathscr{C}} \preceq \mathscr{C}$.

Proof of Lemma 3.3 Obviously, $\widetilde{\mathcal{T}} \subseteq \mathcal{T}$. Apply Lemma 2.7. □

Lemma 3.4 Let $\mathscr{C}$ and $\mathscr{C}'$ be two slim chronicles. If $\mathscr{C} \approx \mathscr{C}'$ then $\mathscr{C} = \mathscr{C}'$.

Proof of Lemma 3.4 This amounts to the antisymmetry property shown in Proposition 3.1. □

Lemma 3.5 If $\mathscr{C} = (\mathcal{E}, \mathcal{T})$ is a chronicle such that $\mathcal{T}$ is finite then $\mathscr{C} \approx \widetilde{\mathscr{C}}$.

Proof of Lemma 3.5 $\widetilde{C} \preceq \mathscr{C}$ is Lemma 3.3. As to the converse, $\mathscr{C} \preceq \widetilde{C}$, consider $(e_i, o)[l : u](e_j, p)$ in $\mathcal{T} \setminus \widetilde{\mathcal{T}}$. From Definition 3.2, there must exist $(e_i, o)[l' : u'](e_j, p) \in \mathcal{T}$ such that $[l' : u'] \subsetneq [l : u]$. Since $\mathcal{T}$ is assumed to be finite, there must also exist $(e_i, o)[l'' : u''](e_j, p)$ in $\mathcal{T}$ such that $[l'' : u''] \subsetneq [l : u]$ and no temporal constraint $(e_i, o)[l''' : u'''](e_j, p)$ in $\mathcal{T}$ satisfy $[l''' : u'''] \subsetneq [l'' : u'']$ hence $(e_i, o)[l'' : u''](e_j, p)$ is in $\widetilde{\mathcal{T}}$. Lastly, $(e_i, o)[l'' : u''](e_j, p) \trianglelefteq (e_i, o)[l : u](e_j, p)$ due to $[l'' : u''] \subsetneq [l : u]$. □

A.3   Proofs for Sect. 3.3

Proposition 3.2 $(\widehat{\mathcal{C}}, \preceq)$ is a poset.

Proof of Proposition 3.2 It is a consequence of Proposition 3.1 $((\widetilde{\mathcal{C}}_f, \preceq)$ is a poset) and Lemma 3.6 $(\widehat{\mathcal{C}} \subset \widetilde{\mathcal{C}}_f)$. □

A.4   Proofs for Sect. 3.5

Lemma 3.7 If $\mathscr{C} = (\mathcal{E}, \mathcal{T})$ and $\mathscr{C}' = (\mathcal{E}', \mathcal{T}')$ are flush then $\mathscr{C}$ and $(\mathcal{E} \Cap \mathcal{E}', \mathcal{T}'')$ are flush.

Proof of Lemma 3.7 Definition 2.7 gives $\mu_{\mathcal{E} \Cap \mathcal{E}'}(e) = \min(\mu_{\mathcal{E}}(e), \mu_{\mathcal{E}'}(e))$ for all $e \in \mathbb{E}$. Let e be such that $\mu_{\mathcal{E} \Cap \mathcal{E}'}(e) \neq 0$. Therefore, $\mu_{\mathcal{E}}(e) \neq 0$ and $\mu_{\mathcal{E}'}(e) \neq 0$. Since $\mathscr{C}$ and $\mathscr{C}'$ are flush, $\mu_{\mathcal{E}}(e) = \mu_{\mathcal{E}'}(e)$ whenever $\mu_{\mathcal{E}}(e) \neq 0$ and $\mu_{\mathcal{E}'}(e) \neq 0$. Hence, $\mu_{\mathcal{E} \Cap \mathcal{E}'}(e) = \mu_{\mathcal{E}}(e)$. □

Lemma 3.8 Let $\mathscr{C} = (\mathcal{E}, \mathcal{T})$ and $\mathscr{C}' = (\mathcal{E}', \mathcal{T}')$ be such that $\mathcal{E}' \Subset \mathcal{E}$. If $\mathscr{C}$ and $\mathscr{C}'$ are flush then there is a single multiset embedding for $\mathcal{E}' \Subset \mathcal{E}$.

Proof of Lemma 3.8 Assume θ_1 and θ_2 to be two multiset embeddings for $\mathcal{E}' \Subset \mathcal{E}$. Consider $e \in \mathbb{E}$ such that $\mu_{\mathcal{E}'}(e) = k > 0$. Then, $e = e'_i = e'_{i+1} = \ldots = e'_{i+k-1}$ for some i (Definition 2.2 requires $\mathcal{E}'$ to be ordered by $\leq_{\mathbb{E}}$). By Definition 2.6, $e'_i = e_{\theta_1(i)} = e_{\theta_1(i+1)} = \ldots = e_{\theta_1(i+k-1)}$ and $e'_i = e_{\theta_2(i)} = e_{\theta_2(i+1)} = \ldots = e_{\theta_2(i+k-1)}$. Since $\mathcal{C}$ and $\mathcal{C}'$ are flush, $\mu_{\mathcal{E}}(e) = k$. Letting h_0 denote the least index such that $e_{h_0} = e$, the fact that θ_1 is strictly increasing (as per Definition 2.6) together with $e = e'_i = e_{\theta_1(i)} = e_{\theta_1(i+1)} = \ldots = e_{\theta_1(i+k-1)}$ and $e \neq e_{\theta_1(j)}$ for $j \notin \{i, \ldots, i + k - 1\}$ (there are exactly k copies of e in $\mathcal{E}$) entail $\theta_1(i) = h_0, \theta_1(i + 1) = h_0 + 1, \ldots, \theta_1(i + k - 1) = h_0 + k - 1$. Of course, the same reasoning holds for θ_2 hence $\theta_2(i) = h_0, \theta_2(i + 1) = h_0 + 1, \ldots, \theta_2(i + k - 1) = h_0 + k - 1$. Therefore, $\theta_1(j) = \theta_2(j)$ for $j \in \{i, \ldots, i + k - 1\}$.

Since this is for each $e \in \mathbb{E}$, it follows that there cannot be more than one multiset embedding for $\mathcal{E} \Cap \mathcal{E}' \Subset \mathcal{E}$. That there exists at least one is given by Definition 2.6.

$\square$

Lemma 3.9 If $\mathcal{C} = (\mathcal{E}, \mathcal{T})$ and $\mathcal{C}' = (\mathcal{E}', \mathcal{T}')$ are flush then there is a unique multiset embedding for $\mathcal{E} \Cap \mathcal{E}' \Subset \mathcal{E}$.

Proof of Lemma 3.9 Apply Lemmas 3.7 and 3.8. $\square$

Lemma 3.10 If $\mathcal{C}$ and $\mathcal{C}'$ are flush then $\mathcal{C} \curlywedge \mathcal{C}' \preceq \mathcal{C}$.

Proof of Lemma 3.10 Consider $(e''_i, i)[l'' : u''](e''_j, j) \in \mathcal{T} \odot \mathcal{T}'$. It follows from Definition 3.4 that $(e''_i, i)[l'' : u''](e''_j, j) = (e''_i, i)[\min(l, l'), \max(u, u')](e''_j, j)$ for some $(e''_i, \theta(i))[l : u](e''_j, \theta(j)) \in \mathcal{T}$ where $\theta \in \Theta$. Since Θ consists of all multisets embeddings for $\mathcal{E} \Cap \mathcal{E}' \Subset \mathcal{E}$, $e''_i = e_{\theta(i)}$ and $e''_j = e_{\theta(j)}$ hence some $(e_{\theta(i)}, \theta(i))[l : u](e_{\theta(j)}, \theta(j))$ is in $\mathcal{T}$. Also, it is clear that $[l : u] \subseteq [\min(l, l'), \max(u, u')]$. Thus, $(e_{\theta(i)}, \theta(i))[l : u](e_{\theta(j)}, \theta(j)) \trianglelefteq (e''_i, i)[l'' : u''](e''_j, j)$.

Lastly, $\preceq$ requires that θ involved in the $\trianglelefteq$ condition be the same when ranging over all $(e''_i, i)[l'' : u''](e''_j, j)$ in $\mathcal{T} \odot \mathcal{T}'$. This is satisfied here because Lemma 3.9 means that $\Theta = \{\theta\}$ since $\mathcal{C}$ and $\mathcal{C}'$ are flush. $\square$

Lemma 3.11 Let $\mathcal{C}, \mathcal{C}', \mathcal{C}''$ be pairwise flush. If $\mathcal{C}''$ is a lower bound of $\mathcal{C}$ and $\mathcal{C}'$ then $\mathcal{C}'' \preceq \mathcal{C} \curlywedge \mathcal{C}'$.

Proof of Lemma 3.11 To make it easier with the notation, we prove that if $\mathcal{C}, \mathcal{C}', \mathcal{C}'''$ are pairwise flush where $\mathcal{C}''' = (\mathcal{E}''', \mathcal{T}''')$ is a lower bound of $\mathcal{C}$ and $\mathcal{C}'$ wrt $\preceq$, then $\mathcal{C}''' \preceq (\mathcal{E} \Cap \mathcal{E}', \mathcal{T} \odot \mathcal{T}')$.

We write $\mathcal{E}''' = \{\!\{e'''_1, \ldots, e'''_p\}\!\}$ and as usual $\mathcal{E} \Cap \mathcal{E}' = \{\!\{e''_1, \ldots, e''_s\}\!\}$.

Since $\mathcal{C}'''$ is a lower bound of $\mathcal{C}$ and $\mathcal{C}'$, Definition 2.8 and Lemma 2.4 entail $\mathcal{E}''' \Subset \mathcal{E} \Cap \mathcal{E}'$.

We are to show that there exists a multiset embedding θ'' for $\mathcal{E}''' \Subset \mathcal{E} \Cap \mathcal{E}'$ such that for every $(e'''_h, h)[l'''_w : u'''_w](e'''_k, k)$ in $\mathcal{T}'''$, $[\min(l, l'), \max(u, u')] \subseteq [l'''_w : u'''_w]$ for some $(e''_{\theta''(h)}, \theta''(h))[\min(l, l'), \max(u, u')](e''_{\theta''(k)}, \theta''(k))$ in $\mathcal{T} \odot \mathcal{T}'$ while $e''_{\theta''(z)} = e'''_z$ for all z.

$\mathcal{C}$ is at least as specific as $\mathcal{C}'''$ hence there is a unique multiset embedding θ_0 for $\{\!\{e'''_1, \ldots, e'''_p\}\!\} = \mathcal{E}''' \Subset \mathcal{E} = \{\!\{e_1, \ldots, e_n\}\!\}$ (Lemma 3.8 applies since $\mathcal{C}$ and $\mathcal{C}'''$

are flush). That is, $\theta_0 : [p] \to [n]$ is strictly increasing and $e_z''' = e_{\theta_0(z)}$ for $z = 1, \ldots, p$ and for every $(e_h''', h)[l''' : u'''](e_k''', k)$ in $\mathcal{T}'''$ there exists $(e_{\theta_0(h)}, \theta_0(h))[l : u](e_{\theta_0(k)}, \theta_0(k))$ in $\mathcal{T}$ such that $e_{\theta_0(h)} = e_h'''$, $e_{\theta_0(k)} = e_k'''$ and $[l : u] \subseteq [l''' : u''']$.

Since $\mathcal{E}$ and $\mathcal{E}'$ are flush, Lemma 3.9 guarantees that there exists a unique multiset embedding θ for $\mathcal{E} \sqcap \mathcal{E}' \Subset \mathcal{E}$ (with $e_{\theta(x)} = e_x''$ for $x = 1, \ldots, s$). Also, it is clear that $\mathcal{E}'''$ and $\mathcal{E} \sqcap \mathcal{E}'$ are flush. Then, Lemma 3.8 ensures that there exists a unique multiset embedding θ'' for $\mathcal{E}''' \Subset \mathcal{E} \sqcap \mathcal{E}'$ (with $e_{\theta''(y)} = e_y'''$ for $y = 1, \ldots, p$). Clearly, $\theta \circ \theta''$ is a multiset embedding for $\mathcal{E}''' \Subset \mathcal{E}$ (since both θ and θ'' are strictly increasing, so is $\theta \circ \theta''$, while $e_y''' = e_{\theta''(y)}$ together with $e_x'' = e_{\theta(x)}$ give $e_{(\theta \circ \theta'')(z)} = e_{\theta(\theta''(z))} = e_{\theta''(z)}'' = e_z'''$). However, θ_0 is already such a multiset embedding. Applying Lemma 3.8, $\theta \circ \theta''$ is θ_0.

Similarly, $\mathcal{C}'$ is at least as specific as $\mathcal{C}'''$ hence the unique multiset embedding θ_0' for $\mathcal{E}''' \Subset \mathcal{E}'$ (Lemma 3.8 applies because $\mathcal{C}'$ and $\mathcal{C}'''$ are flush) is such that for every $(e_h''', h)[l''' : u'''](e_k''', k)$ in $\mathcal{T}'''$ there exists $(e_{\theta_0'(h)}', \theta_0'(h))[l' : u'](e_{\theta_0'(k)}', \theta_0'(k))$ in $\mathcal{T}'$ where $e_{\theta_0'(h)}' = e_h'''$, $e_{\theta_0'(k)}' = e_k'''$ and $[l' : u'] \subseteq [l''' : u''']$.

By the same reasoning as above, there exists a unique multiset embedding θ' for $\mathcal{E} \sqcap \mathcal{E}' \Subset \mathcal{E}'$ and $\theta' \circ \theta''$ is θ_0'.

Consider $(e_h'', h)[l_w''' : u_w'''](e_k'', k)$ in $\mathcal{T}'''$. As was detailed above, $[l : u] \subseteq [l_w''' : u_w''']$ and $[l' : u'] \subseteq [l_w''' : u_w''']$ for some $(e_{\theta_0(h)}, \theta_0(h))[l : u](e_{\theta_0(k)}, \theta_0(k))$ in $\mathcal{T}$ and some $(e_{\theta_0'(h)}', \theta_0'(h))[l' : u'](e_{\theta_0'(k)}', \theta_0'(k))$ in $\mathcal{T}'$ while $e_{\theta_0'(z)}' = e_z''' = e_{\theta(z)}$ for $z = 1, \ldots, p$. Now, $[l : u] \subseteq [l_w''' : u_w''']$ and $[l' : u'] \subseteq [l_w''' : u_w''']$ give $[\min(l, l'), \max(u, u')] \subseteq [l_w''' : u_w''']$. However, $\theta_0(h) = \theta(\theta''(h))$ due to $\theta \circ \theta'' = \theta_0$ and $\theta_0'(h) = \theta'(\theta''(h))$ due to $\theta' \circ \theta'' = \theta_0'$ (the same for k, of course). Applying Definition 3.4 for $i = \theta''(h)$ and $j = \theta''(k)$, it follows that $(e_{\theta(i)}, \theta(i))[l : u](e_{\theta(j)}, \theta(j))$ is in $\mathcal{T}$ and $(e_{\theta'(i)}', \theta'(i))[l' : u'](e_{\theta'(j)}', \theta'(j))$ in $\mathcal{T}'$, giving $(e_i'', i)[\min(l, l'), \max(u, u')](e_j'', j)$ in $\mathcal{T} \odot \mathcal{T}'$. Substituting back to h and k gives the expected $(e_{\theta''(h)}'', \theta''(h))[\min(l, l'), \max(u, u')](e_{\theta''(k)}'', \theta''(k))$ in $\mathcal{T} \odot \mathcal{T}'$. $\quad\square$

Proposition 3.3 $(\widetilde{\mathcal{C}_{fP}}, \preceq)$ is a lower semilattice.

Proof of Proposition 3.3 Trivially, if $\mathcal{C}$ and $\mathcal{C}'$ conform exactly with P then they are flush. By Lemma 3.10, $\mathcal{C} \curlywedge \mathcal{C}'$ is a lower bound of $\mathcal{C}$ and $\mathcal{C}'$ while Lemma 3.11 expresses that $\mathcal{C} \curlywedge \mathcal{C}'$ $\preceq$-dominates every lower bound of $\mathcal{C}$ and $\mathcal{C}'$. Clearly, $\mathcal{C} \curlywedge \mathcal{C}'$ is finite. Applying Lemma 3.5 then gives $\mathcal{C} \curlywedge \mathcal{C}' \approx \widetilde{\mathcal{C} \curlywedge \mathcal{C}'}$, and since two equivalent slim chronicles are identical (Lemma 3.4), $\widetilde{\mathcal{C} \curlywedge \mathcal{C}'}$ is the greatest lower bound of $\mathcal{C}$ and $\mathcal{C}'$. That $\widetilde{\mathcal{C} \curlywedge \mathcal{C}'}$ is in $\widetilde{\mathcal{C}_{fP}}$ follows from Definition 3.2 and Lemma 3.7. $\quad\square$

A.5 Proofs for Sect. 4.2

Lemma 4.2 If $i < j$ then $e_{f(i)} \leq_{\mathbb{E}} e_{f(j)}$.

Proof of Lemma 4.2 Let $\mathscr{C} = (\{\!\{e_1', \ldots, e_m'\}\!\}, \mathcal{T})$ be a chronicle that has an occurrence $\{(e_{f(1)}, t_{f(1)}), \ldots, (e_{f(m)}, t_{f(m)})\}$ in a stripped sequence $S = \langle (e_1, t_1), \ldots, (e_n, t_n) \rangle$. Since $\{\!\{e_1', \ldots, e_m'\}\!\}$ is the multiset of $\mathscr{C}$, Definition 2.2 ensures $e_i' \leq_{\mathbb{E}} e_j'$ for $i < j$. From Definition 4.2(3), $e_i' = e_{f(i)}$ and $e_j' = e_{f(j)}$. Thus, $e_{f(i)} \leq_{\mathbb{E}} e_{f(j)}$ for $i < j$.   □

Proposition 4.1 If a chronicle $\mathscr{C}$ occurs in at least one stripped sequence, then $\mathscr{C}$ is well-behaved.

Proof of Proposition 4.1 Let $S = \langle (e_1, t_1), \ldots, (e_n, t_n) \rangle$ be a stripped sequence in which $\mathscr{C} = (\mathcal{E}, \mathcal{T})$ occurs where $\mathcal{E} = \{\!\{e_1', \ldots, e_m'\}\!\}$. By Definition 4.2, there exists $f : [m] \to [n]$ such that (1) $f : [m] \to [n]$ is an injective function, (2) $f(i) < f(i+1)$ whenever $e_i' = e_{i+1}'$, (3) $e_i' = e_{f(i)}$ for $i = 1, \ldots, m$ (4) $t_{f(j)} - t_{f(i)} \in [t^- : t^+]$ whenever $(e_i', i)[t^- : t^+](e_j', j) \in \mathcal{T}$.

To begin with, observe from Definition 2.2 that if $(e, o)[t^- : t^+](e', o')$ is in $\mathcal{T}^*$ then $e = e_o' \in \mathcal{E}$ and $e' = e_{o'}' \in \mathcal{E}$ hence $(e, o)[t^- : t^+](e', o')$ is in fact of the form $(e_i', i)[t^- : t^+](e_j', j)$ where $i < j$ (and $e_i' \leq_{\mathbb{E}} e_j'$).

For $(e_i', i)[t^- : t^+](e_j', j) \in \mathcal{T}^*$, we show, by structural induction, $e_i' = e_{f(i)}$, $e_j' = e_{f(j)}$ and $t^- \leq t_{f(j)} - t_{f(i)} \leq t^+$.

Consider first $(e_i', i)[t^- : t^+](e_j', j) \in \mathcal{T}$. By Definition 4.2(4), $t_{f(j)} - t_{f(i)} \in [t^- : t^+]$ i.e., $t^- \leq t_{f(j)} - t_{f(i)} \leq t^+$. Besides, Definition 4.2(2) gives $e_i' = e_{f(i)}$ and $e_j' = e_{f(j)}$.

Second, consider $(e_i', i)[\max(t_1^- : t_2^-), \min(t_1^+ : t_2^+)](e_j', j)$ such that all of the following conditions hold: $(e_i', i)[t_1^- : t_1^+](e_j', j)$ and $(e_i', i)[t_2^- : t_2^+](e_j', j)$ are in $\mathcal{T}^*$ and $e_i' = e_{f(i)}$ and $e_j' = e_{f(j)}$ and $t_1^- \leq t_{f(j)} - t_{f(i)} \leq t_1^+$ and $t_2^- \leq t_{f(j)} - t_{f(i)} \leq t_2^+$. Accordingly, $\max(t_1^- : t_2^-) \leq t_{f(j)} - t_{f(i)} \leq \min(t_1^+ : t_2^+)$ while $e_i' = e_{f(i)}$ and $e_j' = e_{f(j)}$.

Third, consider $(e_i', i)[t_1^- + t_2^- : t_1^+ + t_2^+](e_k', k)$ such that $(e_i', i)[t_1^- : t_1^+](e_j', j)$ and $(e_j', j)[t_2^- : t_2^+](e_k', k)$ are in $\mathcal{T}^*$ and $e_i' = e_{f(i)}$ and $e_j' = e_{f(j)}$ and $e_k' = e_{f(k)}$ and $t_1^- \leq t_{f(j)} - t_{f(i)} \leq t_1^+$ and $t_2^- \leq t_{f(k)} - t_{f(j)} \leq t_2^+$. Trivially, $t_1^- + t_2^- \leq t_{f(k)} - t_{f(i)} \leq t_1^+ + t_2^+$ while $e_i' = e_{f(i)}$ and $e_k' = e_{f(k)}$.

Summing it up, for $(e_i', i)[t^- : t^+](e_j', j) \in \mathcal{T}^*$, $e_i' = e_{f(i)}$ and $e_j' = e_{f(j)}$ and $t^- \leq t_{f(j)} - t_{f(i)} \leq t^+$ hence $t^- \leq t^+$.   □

Proposition 4.2 Let $\mathscr{C}$ and $\mathscr{C}'$ be two chronicles such that $\mathscr{C}' \preceq \mathscr{C}$. For all stripped sequence S, if $\mathscr{C}$ occurs in S then $\mathscr{C}'$ occurs in S.

Proof of Proposition 4.2 To make it easier with notations, we actually prove the following statement: Given a stripped sequence $S = \langle (e_1, t_1), \ldots, (e_n, t_n) \rangle$, for two chronicles $\mathscr{C}' = (\{\!\{e_1', \ldots, e_m'\}\!\}, \mathcal{T}')$ and $\mathscr{C}'' = (\{\!\{e_1'', \ldots, e_s''\}\!\}, \mathcal{T}'')$ such that $\mathscr{C}'' \preceq \mathscr{C}'$, if $\mathscr{C}'$ occurs in S then $\mathscr{C}''$ occurs in S.

In view of the definition, the assumption that $\mathscr{C}'$ occurs in S means that there exists a subset $\{(e_{f(1)}, t_{f(1)}), \ldots, (e_{f(m)}, t_{f(m)})\}$ of S such that

(1) $f : [m] \to [n]$ is an injective function,

(2) $f(i) < f(i+1)$ whenever $e'_i = e'_{i+1}$,
(3) $e'_i = e_{f(i)}$ for $i = 1, \ldots, m$ and
(4) $t_{f(j)} - t_{f(i)} \in [t^- : t^+]$ whenever $(e'_i, i)[t^- : t^+](e'_j, j) \in \mathcal{T}'$.

According to Definition 2.8, the assumption $\mathscr{C}'' \preceq \mathscr{C}'$ means that $\mathcal{E}'' \Subset \mathcal{E}$ by some multiset-embedding θ such that for all $(e''_i, o)[l'' : u''](e''_j, p)$ in $\mathcal{T}''$ there exists a temporal constraint $(e'_{\theta(i)}, \theta(o))[l'' : u''](e'_{\theta(j)}, \theta(p))$ in $\mathcal{T}'$ satisfying[1] $(e'_{\theta(i)}, \theta(o))[l'' : u''](e'_{\theta(j)}, \theta(p)) \trianglelefteq (e''_i, o)[l'' : u''](e''_j, p)$.

Also, the assumption that θ is a multiset-embedding for $\mathcal{E}'' \Subset \mathcal{E}'$ means that

(i) $\theta : [s] \to [m]$,
(ii) θ is strictly increasing,
(iii) $e''_i = e'_{\theta(i)}$ for $i = 1, \ldots, s$.

We are to show that for some g,

1. $g : [s] \to [n]$ is an injective function,
2. $g(i) < g(i+1)$ whenever $e''_i = e''_{i+1}$,
3. $e''_i = e_{g(i)}$ for $i = 1, \ldots, s$,
4. $t_{g(j)} - t_{g(i)} \in [t^- : t^+]$ whenever $(e''_i, i)[t^- : t^+](e''_j, j) \in \mathcal{T}''$.

We take $f(\theta(i))$ to be our candidate for $g(i)$.

1. It follows from (1), (i) and (ii) that $f \circ \theta : [s] \to [n]$ is an injective function.
2. Assume $e''_i = e''_{i+1}$. By (iii), $e''_i = e'_{\theta(i)}$ and $e''_{i+1} = e'_{\theta(i+1)}$. In the case $e''_i = e''_{i+1}$, it follows that $e'_{\theta(i)} = e'_{\theta(i+1)}$ hence $e'_{\theta(i)} = e'_{\theta(i)+1} = \cdots = e'_{\theta(i+1)}$ (since θ is strictly increasing and $e'_1 \leq_{\mathbb{E}} \cdots \leq_{\mathbb{E}} e'_m$ by Definition 2.2). Applying (2) iteratively, $f(\theta(i)) < f(\theta(i+1))$.
3. Due to (iii), $e''_i = e'_{\theta(i)}$ for $i = 1, \ldots, s$. Due to (3), $e'_j = e_{f(j)}$ for $j = 1, \ldots, m$. Checking with (i) that the range of θ is $[m]$, $e''_i = e_{f(\theta(i))}$ ensues.
4. Let $(e''_i, i)[t^- : t^+](e''_j, j) \in \mathcal{T}''$. A consequence of the assumption $\mathscr{C}'' \preceq \mathscr{C}'$ is that there exists a temporal constraint $(e'_{\theta(i)}, \theta(i))[l' : u'](e'_{\theta(j)}, \theta(j))$ in $\mathcal{T}'$ satisfying $(e'_{\theta(i)}, \theta(i))[l' : u'](e'_{\theta(j)}, \theta(j)) \trianglelefteq (e''_i, i)[t^- : t^+](e''_j, j)$. Thus, $[l' : u'] \subseteq [t^- : t^+]$. Applying (4), $t_{f(\theta(j))} - t_{f(\theta(i))} \in [l' : u'] \subseteq [t^- : t^+]$.

That is, all four conditions in Definition 4.2 are satisfied for $f \circ \theta$. $\square$

Lemma 4.3 Let S be a stripped sequence. For $1 \leq i < j \leq \sum_{e \in \mathbb{E}} \mu_{\mathcal{E}}(e)$, any occurrence of a chronicle $(\mathcal{E}, \mathcal{T})$ in S is also an occurrence of $(\mathcal{E}, \mathcal{T} \cup \{(e_i, i)[-\infty : +\infty](e_j, j)\})$ in S.

Proof of Lemma 4.3 Once again, we simplify matters by proving the statement for $\mathscr{C}' = (\mathcal{E}' = \{\!\{e'_1, \ldots, e'_m\}\!\}, \mathcal{T}')$ instead of $\mathscr{C}$. We then consider the unlimited temporal constraint $(e'_h, h)[-\infty : +\infty](e'_k, k)$.

---

[1] Since θ is a multiset-embedding for $\mathcal{E}''$ into $\mathcal{E}'$, both $e''_i = e'_{\theta(i)}$ and $e''_j = e'_{\theta(j)}$.

That $\mathscr{C}'$ occurs in $S = \langle (e_1, t_1), \ldots, (e_n, t_n) \rangle$ means that there exists, for some f, a subset $\{(e_{f(1)}, t_{f(1)}), \ldots, (e_{f(m)}, t_{f(m)})\}$ of S such that

(1) $f : [m] \to [n]$ is an injective function,
(2) $f(i) < f(i+1)$ whenever $e_i' = e_{i+1}'$,
(3) $e_i' = e_{f(i)}$ for $i = 1, \ldots, m$,
(4) $t_{f(j)} - t_{f(i)} \in [t^- : t^+]$ whenever $(e_i', i)[t^- : t^+](e_j', j) \in \mathcal{T}'$.

In order to show that $\{(e_{f(1)}, t_{f(1)}), \ldots, (e_{f(m)}, t_{f(m)})\}$ is also an occurrence of $(\mathcal{E}', \mathcal{T}' \cup \{(e_h', h)[-\infty : +\infty](e_k', j)\})$ in S, it must be checked that there exists f satisfying

(1') $f : [m] \to [n]$ is an injective function,
(2') $f(i) < f(i+1)$ whenever $e_i' = e_{i+1}'$,
(3') $e_i' = e_{f(i)}$ for $i = 1, \ldots, m$,
(4') $t_{f(j)} - t_{f(i)} \in [t^- : t^+]$ whenever $(e_i', i)[t^- : t^+](e_j', j) \in \mathcal{T}' \cup \{(e_h', h)[-\infty : +\infty](e_k', k)\}$.

Clearly, the only difference is about $t_{f(k)} - t_{f(h)} \in [-\infty : +\infty]$, something trivial in view of the definition of an event sequence.

So, there is an equivalence between being an occurrence of $(\mathcal{E}', \mathcal{T}')$ in S and being an occurrence of $(\mathcal{E}', \mathcal{T}' \cup \{(e_h', h)[-\infty : +\infty](e_k', k)\})$ in S □

Proposition 4.3 Let S be a sequence, $\mathsf{count}_S(\cdot)$ is anti-monotonic wrt $\preceq$.

Proof of Proposition 4.3 We consider three types of elementary transformations from a chronicle $\mathscr{C}$ to another chronicle $\mathscr{C}'$: adding a node that is not involved in temporal constraints, adding a temporal constraint and narrowing the interval of a temporal constraint. In this three cases, we have $\mathscr{C} \preceq \mathscr{C}'$ (by Definition of $\preceq$).

Whatever the chronicles $\mathscr{C}$ and $\mathscr{C}'$ such that $\mathscr{C} \preceq \mathscr{C}'$, it happens that $\mathscr{C}'$ can be obtained from $\mathscr{C}$ by a sequence of these transformations. This explains why we call them elementary transformations. Then, to show that $\mathsf{count}_S(\cdot)$ is anti-monotonic wrt to $\preceq$, it is sufficient to show that it is anti-monotonic by the three above transformations.

Instead of showing that the measure decreases by one of these transformations, we show that it can not decrease by the reverse one.

Let $\mathscr{C} = (\mathcal{E}, \mathcal{T})$ be a chronicle such that $\mathsf{count}_S(\mathscr{C}) = n$, then, it exists $\mathcal{F} = \{f_1, \ldots, f_n\}$ that are pairwise disjoint embeddings, then for each pair of distinct embeddings, Eq. (4.1) holds.

- Let C' be a chronicle with one less temporal constraint c then, whatever $f \in \mathcal{F}$, $\{(f(i), f(j)) \mid (e_i, i)[l : u](e_j, j) \in \mathcal{T} \setminus \{c\}\} \subset \{(f(i), f(j)) \mid (e_i, i)[l : u](e_j, j) \in \mathcal{T}\}$. Then, for all pairs of distinct embeddings $f, f' \in \mathcal{F}$, $\{(f(i), f(j)) \mid (e_i, i)[l : u](e_j, j) \in \mathcal{T} \setminus \{c\}\}$ remains disjoint from $\{(f'(i), f'(j)) \mid (e_i, i)[l : u](e_j, j) \in \mathcal{T} \cup \{c\}\}$, and thus $\mathcal{F}$ is also a set of n pairwise disjoint embeddings for $\mathscr{C}'$. It ensues that $\mathsf{count}_S(\mathscr{C}') \not< \mathsf{count}_S(\mathscr{C})$, i.e. $\mathsf{count}_S(\mathscr{C}') \geq \mathsf{count}_S(\mathscr{C})$.

- Let $C' = (\mathcal{E}, \mathcal{T}')$ be a chronicle such that it exists u, v such that $\mathcal{T}' = \mathcal{T} \setminus \{(e_u, u)[t^- : t^+](e_v, v)\} \cup \{(e_u, u)[t'^- : t'^+](e_v, v)\}$ with $[t'^- : t'^+] \supset [t^- : t^+]$ then for all $f \in \mathcal{F}$, f is also an embedding of $\mathscr{C}'$ so $\mathcal{F}$ is a set n disjoint embeddings of $\mathscr{C}'$. It ensues that $\mathsf{count}_S(\mathscr{C}') \not< \mathsf{count}_S(\mathscr{C})$, i.e. $\mathsf{count}_S(\mathscr{C}') \geq \mathsf{count}_S(\mathscr{C})$.

- Let $C' = (\mathcal{E}', \mathcal{T}')$ be a chronicle such that $\mathcal{E}' = \mathcal{E} \setminus \{\!\{e\}\!\}$ where e is not involved in $\mathcal{T}$. We denote by θ the multiset embedding of $\mathcal{E}'$ in $\mathcal{E}$ and $\mathcal{T}' = \{(e_{\theta(i)}, \theta(i))[l : u](e_{\theta(j)}, \theta(j)) \mid (e_i, i)[l : u](e_j, j) \in \mathcal{T}\}$.

  For all $f \in \mathcal{F}$, f' is defined by $f'(i) = f(\theta(i))$ for all $i \in [m-1]$. It ensues that f' is an embedding of $\mathscr{C}'$ in S. Then, $\{f'_1, \ldots, f'_n\}$ is a collection n embeddings of $\mathscr{C}'$

  We can notice that, by construction, $\{(f'(i), f'(j)) \mid (e_i, i)[l : u](e_j, j) \in \mathcal{T}'\} = \{(f(i), f(j)) \mid (e_i, i)[l : u](e_j, j) \in \mathcal{T}\}$. This means that these sets are pairwise disjoints. Then again, we can conclude that necessarily $\mathsf{count}_S(\mathscr{C}') \geq \mathsf{count}_S(\mathscr{C})$.

$\square$

A.6 Proofs for Sect. 5.2

Lemma 5.1 φ in Definition 5.1 is unique.

Proof of Lemma 5.1 Assume that φ_1 and φ_2 are such that $e_{\varphi_w(i)} \leq_\mathbb{E} e_{\varphi_w(i+1)}$ and if $e_{\varphi_w(i)} = e_{\varphi_w(i+1)}$ then $\varphi_w(i) < \varphi_w(i+1)$ for $w = 1, 2$. Hence, $e_{\varphi_1(1)} \leq_\mathbb{E} e_j$ and $e_{\varphi_2(1)} \leq_\mathbb{E} e_j$ for $j = 1, \ldots, n$ since φ_1 and φ_2 are permutations over $[n]$. Then, $e_{\varphi_1(1)} \leq_\mathbb{E} e_{\varphi_2(1)}$ and $e_{\varphi_2(1)} \leq_\mathbb{E} e_{\varphi_1(1)}$. Thus, $e_{\varphi_1(1)} = e_{\varphi_2(1)}$. However, $\varphi_2(1)$ is $\varphi_1(h)$ for some h. Also, $e_{\varphi_1(i)} = e_{\varphi_1(i+1)}$ implies $\varphi_1(i) < \varphi_1(i+1)$. Hence $e_{\varphi_1(1)} = e_{\varphi_1(h)}$ gives $\varphi_1(1) < \varphi_1(h) = \varphi_2(1)$ unless $h = 1$. Similarly, $\varphi_1(1)$ is $\varphi_2(k)$ for some k and it follows that $\varphi_2(1) < \varphi_2(k) = \varphi_1(1)$ unless $k = 1$. Since $\varphi_1(1) < \varphi_2(1)$ and $\varphi_2(1) < \varphi_1(1)$ cannot hold together, either $h = 1$ or $k = 1$. Each case gives $\varphi_1(1) = \varphi_2(1)$. The same reasoning can in turn be applied to the restriction of $\varphi_1 : \{2, \ldots, n\} \to [n] \setminus \{\varphi_1(1)\}$ (similarly for φ_2) then to $\varphi_1 : \{3, \ldots, n\} \to [n] \setminus \{\varphi_1(1), \varphi_1(2)\}$ (similarly for φ_2) and so on. $\square$

Lemma 5.3 $\delta(S)$ is a slim chronicle.

Proof of Lemma 5.3 It is obvious that $\delta(S)$ is slim. It is routine to check that $\delta(S)$ is a chronicle: in particular, the condition $t^-, t^+ \in \overline{\mathbb{T}}$ in Definition 2.2 is met in the form $t_{\varphi(j)} - t_{\varphi(i)} \in \mathbb{T}$ that holds due to the requirement in Definition 2.1 for $\mathbb{T}$ to be closed under substraction. $\square$

Proposition 5.1 $\delta(S)$ occurs in S.

Proof of Proposition 5.1 According to Definition 5.1, $\mathcal{E}_{\delta(S)} = \{\!\{e_{\varphi(1)}, \ldots, e_{\varphi(n)}\}\!\}$. Define $\widetilde{S} = \{(e_{\varphi(1)}, t_{\varphi(1)}), \ldots, (e_{\varphi(n)}, t_{\varphi(n)})\}$. We are to show that φ is an f satisfying conditions 1–4 of Definition 4.2 (letting e'_i to be $e_{\varphi(i)}$).

[Condition 1] As a consequence of the fact that φ is a permutation, φ is injective. [Condition 2] Definition 5.1 requires φ to be such that $\varphi(i) < \varphi(i + 1)$ if $e_{\varphi(i)} = e_{\varphi(i+1)}$. [Condition 3] By $\mathcal{E}_{\delta(S)} = \{\!\{e_{\varphi(1)}, \ldots, e_{\varphi(n)}\}\!\}$, it happens that $e'_i = e_{f(i)}$ for f being φ. [Condition 4] Due to Definition 5.1, each temporal constraint in $\mathcal{T}_{\delta(S)}$ is of the form $(e_{\varphi(i)}, i)[t : t](e_{\varphi(j)}, j)$ where $t = t_{\varphi(j)} - t_{\varphi(i)}$. Trivially, $t_{\varphi(j)} - t_{\varphi(i)} \in [t : t]$. □

Proposition 5.2 $\mathscr{C} \preceq \delta(S)$ iff $\mathscr{C}$ occurs in S.

Proof of Proposition 5.2 We start by proving that if $\mathscr{C}$ occurs in S then $\mathscr{C} \preceq \delta(S)$.

Let $\mathscr{C} = (\{\!\{e'_1, \ldots, e'_m\}\!\}, \mathcal{T})$. According to the definition, $\mathscr{C} \preceq \delta(S)$ iff $\{\!\{e'_1, \ldots, e'_m\}\!\} \in \mathcal{E}_{\delta(S)} = \{\!\{e''_1, \ldots, e''_n\}\!\}$ by some multiset embedding θ (i.e., $\theta : [m] \to [n]$ is strictly increasing and $e'_i = e''_{\theta(i)}$ for $i = 1, \ldots, m$) such that for all $(e'_i, i)[l' : u'](e'_j, j)$ in $\mathcal{T}$, there exists $(e''_{\theta(i)}, \theta(i))[l'' : u''](e''_{\theta(j)}, \theta(j))$ in $\mathcal{T}_{\delta(S)}$ satisfying[2] $(e''_{\theta(i)}, \theta(i))[l'' : u''](e''_{\theta(j)}, \theta(j)) \trianglelefteq (e'_i, i)[l' : u'](e'_j, j)$, that is, $e''_{\theta(i)} = e'_i$, $e''_{\theta(j)} = e'_j$ and $[l'' : u''] \subseteq [l' : u']$. The proof consists of finding such a θ.

Since $\mathscr{C}$ occurs in S, there exists (by Definition 4.2) an injection $f : [m] \to [n]$ such that

- $f(i) < f(i + 1)$ whenever $e'_i = e'_{i+1}$,
- $e'_i = e_{f(i)}$ for $i = 1, \ldots, m$,
- $t_{f(j)} - t_{f(i)} \in [t^- : t^+]$ whenever $(e'_i, i)[t^- : t^+](e'_j, j) \in \mathcal{T}$.

We take $\theta(i)$ to be $\varphi^{-1}(f(i))$ which is well-defined since φ is a permutation on $[n]$. We show that $\{\!\{e'_1, \ldots, e'_m\}\!\} \in \mathcal{E}_{\delta(S)} = \{\!\{e_{\varphi(1)}, \ldots, e_{\varphi(n)}\}\!\}$ by virtue of θ. Definition 2.6 requires two conditions.

The first condition is that $e'_i = e''_{\theta(i)}$ must be shown. Since $\{\!\{e_{\varphi(1)}, \ldots, e_{\varphi(n)}\}\!\}$ and $\{\!\{e''_1, \ldots, e''_n\}\!\}$ both denote $\mathcal{E}_{\delta(S)}$ ordered by $\leq_E$, it happens that $e''_j = e_{\varphi(j)}$ hence $e''_{\theta(i)} = e_{\varphi(\theta(i))}$. However, $e_{\varphi(\theta(i))} = e_{\varphi(\varphi^{-1}(f(i)))} = e_{f(i)}$. In view of Definition 4.2(3), $e_{f(i)} = e'_i$ hence $e''_{\theta(i)} = e'_i$ ensues.

The second condition in Definition 2.6 requires that θ must be shown to be strictly increasing. Let $h = \varphi^{-1}(f(i))$ and $k = \varphi^{-1}(f(i+1))$. That is, $\varphi(h) = f(i)$ and $\varphi(k) = f(i + 1)$. Due to Definition 4.2(3), $e'_i = e_{f(i)}$ and $e'_{i+1} = e_{f(i+1)}$. Then, $e'_i = e_{f(i)} = e_{\varphi(h)}$ and $e'_{i+1} = e_{f(i+1)} = e_{\varphi(k)}$. Since C is a chronicle, $e'_i \leq_E e'_{i+1}$. Assume $e'_i <_E e'_{i+1}$. Clearly, $e_{\varphi(h)} <_E e_{\varphi(k)}$. However, Definition 5.1 imposes that $e_{\varphi(j)} \leq_E e_{\varphi(l)}$ for $1 \leq j \leq l \leq n$. Since $\leq_E$ is linear, $e_{\varphi(l)} <_E e_{\varphi(j)}$ thus entails $l < j$. Then, $h < k$. That is, $\theta(i) < \theta(i + 1)$. Otherwise, assume $e'_i = e'_{i+1}$ (the remaining case after $e'_i <_E e'_{i+1}$). Thus, $e_{\varphi(h)} = e_{\varphi(k)}$. By Definition 4.2(2), $f(i) < f(i + 1)$. Since $\varphi(h) = f(i)$ and $\varphi(k) = f(i + 1)$, it follows that $\varphi(h) < \varphi(k)$. Now, the second condition on φ in Definition 5.1 can be rewritten as follows: if $e_{\varphi(l)} = e_{\varphi(j)}$ and $j < l$ then $\varphi(j) < \varphi(l)$. Contrapositively, if $e_{\varphi(l)} = e_{\varphi(j)}$ and $\varphi(l) \leq \varphi(j)$ then $l \leq j$. As we have shown $e_{\varphi(h)} = e_{\varphi(k)}$ and

---

[2] Since θ is a multiset embedding for $\mathcal{E}'$ into $\mathcal{E}$, both $e'_i = e_{\theta(i)}$ and $e'_j = e_{\theta(j)}$.

$\varphi(h) < \varphi(k)$, this obviously gives $h \leq k$ hence $h < k$ (because $\varphi(h) < \varphi(k)$ can only hold for $h \neq k$). That is, $\theta(i) < \theta(i+1)$.

We have shown $\{\{e'_1, \ldots, e'_m\}\} \in \mathcal{E}_{\delta(S)}$. From Definition 2.8, there remains to show that each temporal constraint in $\mathcal{T}$ is $\trianglelefteq$-dominated by a temporal constraint in $\mathcal{T}_{\delta(S)}$. Consider $(e'_i, i)[l' : u'](e'_j, j)$ in $\mathcal{T}$. By Definition 2.2, $i < j$. Definition 4.2(3) gives $e'_i = e_{f(i)}$ and $e'_j = e_{f(j)}$. Thus, $(e'_i, i)[l' : u'](e'_j, j)$ is $(e_{f(i)}, i)[l' : u'](e_{f(j)}, j)$. Definition 4.2(4) gives $t_{f(j)} - t_{f(i)} \in [l' : u']$. Let $h = \varphi^{-1}(f(i))$ and $k = \varphi^{-1}(f(j))$. Therefore, $\varphi(h) = f(i)$ and $\varphi(k) = f(j)$. Consequently, $(e'_i, i)[l' : u'](e'_j, j)$ is $(e_{\varphi(h)}, i)[l' : u'](e_{\varphi(k)}, j)$ and $t_{\varphi(k)} - t_{\varphi(h)} \in [l' : u']$. Moreover, $i < j$ entails $h < k$ because θ has just been shown to be strictly increasing ($h = \theta(i)$ and $k = \theta(j)$). Definition 5.1 guarantees that there exists in $\mathcal{T}_{\delta(S)}$ at least one temporal constraint $(e_{\varphi(h)}, h)[t : t](e_{\varphi(k)}, k)$ for $t = t_{\varphi(k)} - t_{\varphi(h)}$. However, $e''_h = e_{\varphi(h)}$ and $e''_k = e_{\varphi(k)}$ since $\{\{e_{\varphi(1)}, \ldots, e_{\varphi(n)}\}\}$ and $\{\{e''_1, \ldots, e''_n\}\}$ both denote $\mathcal{E}_{\delta(S)}$ ordered by $\leq_{\mathbb{E}}$, hence $(e_{\varphi(h)}, h)[t : t](e_{\varphi(k)}, k) = (e''_h, h)[t : t](e''_k, k)$. Since $h = \theta(i)$ and $k = \theta(j)$, it follows that $(e_{\varphi(h)}, h)[t : t](e_{\varphi(k)}, k) = (e''_{\theta(i)}, \theta(i))[t : t](e''_{\theta(j)}, \theta(j))$. Moreover, $[t : t] \subseteq [l' : u']$ because $t = t_{\varphi(k)} - t_{\varphi(h)}$ while $t_{f(j)} - t_{f(i)} \in [l' : u']$. To sum up, for $(e'_i, i)[l' : u'](e'_j, j)$ in $\mathcal{T}$, there exists $(e''_{\theta(i)}, \theta(i))[t : t](e''_{\theta(j)}, \theta(j))$ in $\mathcal{T}_{\delta(S)}$ such that $e'_i = e''_{\theta(i)}$ and $e'_j = e''_{\theta(j)}$ and $[t : t] \subseteq [l' : u']$, that is, $(e''_{\theta(i)}, \theta(i))[t : t](e''_{\theta(j)}, \theta(j)) \trianglelefteq (e'_i, i)[l' : u'](e'_j, j)$ as required by Definition 2.8.

As to the converse, an instance of Proposition 4.2 is that if $\mathscr{C} \preceq \delta(S)$ and $\delta(S)$ occurs in S then $\mathscr{C}$ occurs in S. However, $\delta(S)$ occurs in S holds by Proposition 5.1. Therefore, if $\mathscr{C} \preceq \delta(S)$ then $\mathscr{C}$ occurs in S. $\qquad\square$

A.7 Proofs for Sect. 5.3.3

Lemma 5.7 If the domain of Λ_μ is restricted to the pure chronicles defined over μ and $\mathbb{T}$, Λ_μ is injective.

Proof of Lemma 5.7 Let $\mathscr{C}_1 = (\mathcal{E}_1, \mathcal{T}_1)$ and $\mathscr{C}_2 = (\mathcal{E}_2, \mathcal{T}_2)$ be two distinct pure chronicles over $\mathbb{E}_D$ and $\mathbb{T}$. Let us write $\mathcal{E}_1 = \{\{e'_{1,1}, \ldots, e'_{1,m_1}\}\}$ and $\mathcal{E}_2 = \{\{e'_{2,1}, \ldots, e'_{2,m_2}\}\}$. Also, we write $\mathcal{E}_D = \{\{e_1, \ldots, e_n\}\}$ as above.

Assume $\Lambda_D(\mathscr{C}_1) = \Lambda_D(\mathscr{C}_2)$. This means $\widetilde{\Lambda_{\mathcal{E}_1}(\mathcal{T}_1)} = \widetilde{\Lambda_{\mathcal{E}_2}(\mathcal{T}_2)}$.

- Consider first the case that $\mathcal{E}_1 \neq \mathcal{E}_2$. Accordingly, there exists $e \in \mathbb{E}$ such that $\mu_{\mathcal{E}_1}(e) \neq \mu_{\mathcal{E}_2}(e)$, i.e., either $\mu_{\mathcal{E}_1}(e)$ is greater or $\mu_{\mathcal{E}_2}(e)$ is greater. The two cases are symmetric, it is enough to deal with $\mu_{\mathcal{E}_2}(e) < \mu_{\mathcal{E}_1}(e) = k$ for some k. There thus exists i such that $e = e'_{1,i-k+1} = \cdots = e'_{1,i}$ but there is no such sequence in $e'_{2,1}, \ldots, e'_{2,m_2}$ (i.e. $\lambda_{\mathcal{E}_1}(i) \notin \operatorname{Im} \lambda_{\mathcal{E}_2}$). To meet the condition $\sum_{e \in \mathbb{E}} \mu_{\mathcal{E}_1}(e) > 1$ from Definition 5.4, there must exist $j \neq i$ such that $e'_j \in \mathcal{E}_1$. We assume $i < j$, it will be clear that the case $j > i$ can be handled similarly. Then, there is some $(e_{\lambda_{\mathcal{E}_1}(i)}, \lambda_{\mathcal{E}_1}(i))[l : u](e_{\lambda_{\mathcal{E}_1}(j)}, \lambda_{\mathcal{E}_1}(j))$ in $\Lambda_{\mathcal{E}_1}(\mathcal{T}_1)$. However, $\lambda_{\mathcal{E}_1}(i) \notin \operatorname{Im} \lambda_{\mathcal{E}_2}$

hence this temporal constraint cannot be in $\Lambda_{\mathcal{E}_2}(\mathcal{T}_2)$, a contradiction (since $\widetilde{\Lambda_{\mathcal{E}_1}(\mathcal{T}_1)}$ and $\widetilde{\Lambda_{\mathcal{E}_2}(\mathcal{T}_2)}$ are obtained from $\Lambda_{\mathcal{E}_1}(\mathcal{T}_1)$ and $\Lambda_{\mathcal{E}_2}(\mathcal{T}_2)$ by slimming).

- Consider now the case that $\mathcal{E}_1 = \mathcal{E}_2$. For $\mathscr{C}_1$ and $\mathscr{C}_2$ to be distinct, it must then be that $\mathcal{T}_1 \neq \mathcal{T}_2$ i.e. either $\mathcal{T}_1 \nsubseteq \mathcal{T}_2$ or $\mathcal{T}_2 \nsubseteq \mathcal{T}_1$. The two cases are symmetric, it is enough to deal with the latter. Then, there exists some $(e'_i, i)[l : u](e'_j, j) \in \mathcal{T}_2 \setminus \mathcal{T}_1$. However, $\mathcal{E}_1 = \mathcal{E}_2$ entails $\lambda_{\mathcal{E}_1} = \lambda_{\mathcal{E}_2}$. Since neither $\mathcal{T}_1$ nor $\mathcal{T}_2$ have a temporal constraint of the form $(e'_i, i)[-\infty : +\infty](e'_j, j)$, it happens that $\lambda_{\mathcal{E}_1} = \lambda_{\mathcal{E}_2}$ reduces $\Lambda_{\mathcal{E}_1}(\mathcal{T}_1) = \Lambda_{\mathcal{E}_2}(\mathcal{T}_2)$ to

$$\{(e_{\lambda_{\mathcal{E}_1}(i)}, \lambda_{\mathcal{E}_1}(i))[l : u](e_{\lambda_{\mathcal{E}_1}(j)}, \lambda_{\mathcal{E}_1}(j)) \mid (e'_i, i)[l : u](e'_j, j) \in \mathcal{T}_1\}$$
$$= \{(e_{\lambda_{\mathcal{E}_2}(i)}, \lambda_{\mathcal{E}_2}(i))[l : u](e_{\lambda_{\mathcal{E}_2}(j)}, \lambda_{\mathcal{E}_2}(j)) \mid (e'_i, i)[l : u](e'_j, j) \in \mathcal{T}_2\}.$$

Then, it ensues that for each $(e'_i, i)[l : u](e'_j, j) \in \mathcal{T}_2 \setminus \mathcal{T}_1$, there exists $(e'_i, i')[l : u](e'_j, j') \in \mathcal{T}_1$, $(i', j') \neq (i, j)$ such that $(e'_{\lambda_{\mathcal{E}_2}(i')}, \lambda_{\mathcal{E}_2}(i'))[l : u](e'_{\lambda_{\mathcal{E}_2}(j')}, \lambda_{\mathcal{E}_2}(j')) = (e'_{\lambda_{\mathcal{E}_2}(i)}, \lambda_{\mathcal{E}_2}(i))[l : u](e'_{\lambda_{\mathcal{E}_2}(j)}, \lambda_{\mathcal{E}_2}(j))$ (otherwise, the sets above would not be equals). But, $\lambda_{\mathcal{E}_2}$ is injective by construction (see Definition 5.3). Thus, there is a contradiction.

□

Lemma 5.8 Let $S = \langle (e_1, t_1), \ldots, (e_m, t_m) \rangle$ be a sequence conforming with a bounded profile P, then the chronicle $\Lambda_P(\delta(S)) \overset{\text{def}}{=} (P, \mathcal{T})$ is such that

$$\mathcal{T} = \{(e_{\lambda(i)}, \lambda(i))[d : d](e_{\lambda(j)}, \lambda(j)) \mid 1 \leq i < j \leq m\}$$

and λ selects a subset of event types in P.

Proof of Lemma 5.8 Let $S = \langle (e_1, t_1), \ldots (e_m, t_m) \rangle$ and P be a profile, following the notations of Definition 5.3, we denote $\delta(S) = (\mathcal{E}, \mathcal{T})$ a chronicle conformed to P where $\mathcal{E} = \{\!\{e'_1, \ldots, e'_m\}\!\}$ and $\lambda_{\mathcal{E}}$ the embedding of $\mathcal{E}$ in P. We remind that there is a permutation (φ) between the sequence of e_i and the sequence of e'_i to ensure the correct ordering of the event types in the multiset $\mathcal{E}$ (see Definition 5.1). We denote $d_{ij} = t_{\varphi(j)} - t_{\varphi(i)}$ the delay between e'_i and e'_j.

Then, by Definition 5.1, we have

$$\mathcal{T} = \left\{ (e'_i, i)[d_{ij} : d_{ij}](e'_j, j) \mid 1 \leq i < j \leq m \right\}$$

By definition of a reduct, we have $\Lambda_P(\delta(S)) = (P, \widetilde{\Lambda_{\mathcal{E}}(\mathcal{T})})$. Applying the definition of $\Lambda_{\mathcal{E}}$ on the set of temporal constraints, it comes:

$$\Lambda_{\mathcal{E}}(\mathcal{T}) = \left\{ (e'_{\lambda_{\mathcal{E}}(i)}, \lambda_{\mathcal{E}}(i))[d_{ij} : d_{ij}](e'_{\lambda_{\mathcal{E}}(j)}, \lambda_{\mathcal{E}}(j)) \mid 1 \leq i < j \leq m \right\} \cup$$
$$\left\{ (e'_{\lambda_{\mathcal{E}}(i)}, \lambda_{\mathcal{E}}(i))[-\infty : +\infty](e'_{\lambda_{\mathcal{E}}(j)}, \lambda_{\mathcal{E}}(j)) \mid 1 \leq i < j \leq m \right\}$$

Obviously, the sliming operation removes each temporal constraint of the second set for which there is a smaller ($\trianglelefteq$) temporal constraint in the first set. Then, we obtain the result of the Lemma. $\qquad\square$

Lemma 5.12 Let μ be a profile and S be a sequence s.t. $\delta(S)$ is conformed to μ, the $\Lambda_\mu\,(\delta\,(S))$ is simple pure.

Proof of Lemma 5.12 Let μ be a profile and $S = \langle(e_1, t_1), \ldots, (e_n, t_n)\rangle$ such that $\delta(S) = (\mathcal{E}, \mathcal{T})$ is conformed to μ,

- if $n \leq 1$ then $\mathcal{T}$ is empty and $\widetilde{\Lambda_\mu(\mathcal{T})}$ is also empty, then $\Lambda_\mu\,(\delta\,(S))$ is simple pure
- $n > 1$ then where $\mathcal{E} = \{\{e_1, \ldots, e_n\}\} \Subset \mu$ and $\mathcal{T} = \{(e_i, i)[d_{ij} : d_{ij}](e_j, j) \mid 1 \leq i < j \leq n, d_{ij} = t_j - t_i\}$. $\lambda_\mathcal{E}$ denotes the embedding of $\mathcal{E}$ in μ. As there is a temporal constraint between the nodes of each pair of $\delta(S)$, then all the infinite temporal constraints are removed by the sliming operator, and then
$$\widetilde{\Lambda_\mu(\mathcal{T})} = \{(e_{\lambda_\mathcal{E}(i)}, \lambda_\mathcal{E}(i))[l : u](e_{\lambda_\mathcal{E}(j)}, \lambda_\mathcal{E}(j)) \mid (e_i', i)[l : u](e_j', j) \in \mathcal{T}\}.$$
Then, $\widetilde{\Lambda_\mu(\delta(S))}$ is simple and pure.

$\qquad\square$

Lemma 5.13 Let $\mathscr{C} = (\mathcal{E}'', \mathcal{T})$ and $\mathscr{C}' = (\mathcal{E}'', \mathcal{T}')$ be two simple pure chronicles sharing the same multiset $\mathcal{E}''$, then $\mathscr{C} \curlywedge \mathscr{C}'$ is simple pure.

Proof of Lemma 5.13 Let $\mathscr{C} = (\mathcal{E}'', \mathcal{T})$ and $\mathscr{C}' = (\mathcal{E}'', \mathcal{T}')$ be two simple pure chronicles sharing the same multiset $\mathcal{E}'' = \{\{e_1'', \ldots, e_n''\}\}$, then in the intersection $\Theta = \{Id\}$ and $\Theta' = \{Id\}$. So, $\mathcal{T} \odot \mathcal{T}'$ is simply

$$\{(e_i'', i)[min(l, l') : max(u, u')](e_j'', j) \mid 1 \leq i < j \leq n,$$

$$(e_i'', i)[l : u](e_j'', j) \in \mathcal{T},$$

$$(e_i'', i)[l' : u'](e_j'', j) \in \mathcal{T}'\}$$

And then, obviously

- $\mathcal{E}''$ is finite,
- $\mathcal{T} \odot \mathcal{T}'$ is simple because $\mathscr{C}$ and $\mathscr{C}$ are simple
- $\mathcal{T} \odot \mathcal{T}'$ does not hold any infinite temporal constraints because $\mathscr{C}$ and $\mathscr{C}$ are pure.

and then $\mathscr{C} \curlywedge \mathscr{C}' = (\mathcal{E}'', \mathcal{T} \odot \mathcal{T}')$ is a simple pure chronicle. $\qquad\square$

Lemma 5.14 glb $\Lambda_{\mathcal{E}_\mathcal{D}}(\mathcal{D})$ is simple pure.

Proof of Lemma 5.14 For all $S \in \mathcal{D}$, $\Lambda_{\mathcal{E}_\mathcal{D}}\,(\delta\,(S))$ is a simple pure chronicle according to Lemma 5.12, and because $\delta(S)$ is conformed to $\mu_\mathcal{D}$ (Lemma 5.10). Then, there chronicles are finite and slim by Definition 5.4, and all conform exactly to $\mu_\mathcal{D}$. Then, there all all in the lattice $\widetilde{\mathcal{C}_{fP}}$ (Prop. 3.3) with $P = \mu_\mathcal{D}$ such that glb $\Lambda_{\mathcal{E}_\mathcal{D}}\,(\delta\,(S))$ exists and is unique. More specifically, glb $\Lambda_{\mathcal{E}_\mathcal{D}}\,(\delta\,(S)) =$

$\bigwedge_{S \in \mathcal{D}} \Lambda_{\mathcal{E}_{\mathcal{D}}} (\delta (S))$. Then, according the Lemma 5.12, glb $\Lambda_{\mathcal{E}_{\mathcal{D}}} (\delta (S))$ is finite pure.

□

Lemma 5.16 There is a unique multiset embedding from $\overline{\mathcal{E}}$ to $\mathcal{E}$.

Proof of Lemma 5.16 A multiset embedding must be strictly increasing, and the multiplicity functions of $\overline{\mathcal{E}}$ and $\{\!\{e_i \in \mathcal{E} \mid \nexists(e_i, i)[l : u](e_j, j) \in \mathcal{T} \wedge \nexists(e_j, j)[l : u](e_i, i) \in \mathcal{T}, \ j \in [n]\}\!\}$ are the same.

□

Lemma 5.17 $\mathcal{E}_{abs(\mathcal{D})} \Subset \mathcal{E}_{\delta(S)}$, for all $S \in \mathcal{D}$

Proof of Lemma 5.17 In general $\mathcal{E}_{\mathcal{C}}$ (resp. $\mathcal{T}_{\mathcal{C}}$) denotes the multiset (resp. the temporal constraints) of a chronicle $\mathcal{C}$ and $\mu_{\mathcal{E}}$ denotes the multiplicity function of a multiset $\mathcal{E}$.

Let start by proving the following intermediary result:

$$(e_i, i)[l : u](e_j, j) \in \mathcal{T}_{\mathcal{D}} \implies \forall S \in \mathcal{D}, \exists(e_i, i)[d : d](e_j, j) \in \mathcal{T}_{\Lambda_{\mathcal{E}_{\mathcal{D}}}(\mathcal{E}_{\delta(S)})},$$
$$s.t. \, d \in [l : u]$$

(A.1)

It is worth notice that the indices of the temporal constraints are the same in the left part and in the right part of the Eq. (A.1).

By Definition 3.4, if $\mathcal{C} = (\mathcal{E}, \mathcal{T})$ and $\mathcal{C}' = (\mathcal{E}', \mathcal{T}')$ conform exactly to μ then $\mathcal{C} \curlywedge \mathcal{C}' = (\mathcal{E} \Cap \mathcal{E}', \mathcal{T}'')$ conforms exactly to μ and then $\Theta = \{Id\}$ and $\Theta' = \{Id\}$. It comes easily that:

$$(e_i, i)[l'' : u''](e_j, j) \in \mathcal{T}'' \implies \exists(e_i, i)[l : u](e_j, j) \in \mathcal{T}, \ [l : u] \subseteq [l'' : u''] \wedge$$
$$\exists(e_i, i)[l' : u'](e_j, j) \in \mathcal{T}', \ [l' : u'] \subseteq [l'' : u'']$$

(A.2)

glb $\Lambda_{\mathcal{E}_{\mathcal{D}}}(\mathcal{D})$ is a min in the semi-lattice of chronicles conformed to $\Lambda_{\mathcal{E}_{\mathcal{D}}}$, and for all $S \in \mathcal{D}$, $\Lambda_{\mathcal{E}_{\mathcal{D}}}(\delta(S))$ lies in this semi-lattice. Then (general property of a glb), for each $S \in \mathcal{D}$:

$$\text{glb } \Lambda_{\mathcal{E}_{\mathcal{D}}}(\mathcal{D}) \curlywedge \Lambda_{\mathcal{E}_{\mathcal{D}}}(\delta(S)) = \text{glb } \Lambda_{\mathcal{E}_{\mathcal{D}}}(\mathcal{D})$$

So, we can deduce from Eq. (A.2) that:

$$(e_i, i)[l'' : u''](e_j, j) \in \mathcal{T}_{\mathcal{D}} \implies \forall S \in \mathcal{D}, \exists(e_i, i)[l : u](e_j, j) \in \mathcal{T}_{\Lambda_{\mathcal{E}_{\mathcal{D}}}(\delta(S))},$$
$$[l : u] \subseteq [l'' : u'']$$

This proves the Eq. (A.1).

We now come back to the proof of the main result. Let $\mathcal{D}$ be a dataset and $S \in \mathcal{D}$ a sequence.

We denote:

- glb $\Lambda_{\mathcal{E}_D}(\mathcal{D}) = (\mathcal{E}_D, \mathcal{T}_D)$ where $\mathcal{E}_D = \{\!\{e'_1, \ldots, e'_n\}\!\}$
- $abs(\mathcal{D}) = \overline{\text{glb } \Lambda_{\mathcal{E}_D}(\mathcal{D})} = (\overline{\mathcal{E}_D}, \overline{\mathcal{T}_D})$, where $\overline{\mathcal{E}_D} = \{\!\{e''_1, \ldots, e''_{n'}\}\!\}$
- $\delta(S) = (\mathcal{E}_S, \mathcal{T}_S)$, where $\mathcal{E}_S = \{\!\{e^S_{\varphi(1)}, \ldots, e^S_{\varphi(m)}\}\!\}$. φ is the bijective reordering function of the event types of S satisfying the multiset order.

Let us proceed by contradiction, and then assume that $\exists S \in D$ s.t. $\overline{\mathcal{E}_D} \not\subseteq \mathcal{E}_S$ then

$$\exists e \in \mathbb{E}, \text{ s.t. } \mu_{\overline{\mathcal{E}_D}}(e) > \mu_{\delta(S)}(e) \tag{A.3}$$

Let us denote i_e (resp. i_1) the index of the last (resp. first) element of type e in $\overline{\mathcal{E}_D}$, then $i_e - i_1 + 1 = \mu_{\overline{\mathcal{E}_D}}(e)$.

By Definition 5.5, it exists j s.t. there is a temporal constraint involving both e'_j and e'_{i_e} (otherwise, e'_{i_e} would have been removed). A similar temporal constraint must also exist between $e''_{\lambda(j)}$ and $e''_{\lambda(i_e)}$ in glb $\Lambda_{\mathcal{E}_D}(\mathcal{D})$ where λ is an embedding of $\overline{\mathcal{E}_D}$ into $\mathcal{E}_D$.

So, by Eq. (A.1), it exists a temporal constraint between $e''_{\lambda(j)}$ and $e''_{\lambda(i_e)}$ in $\Lambda_{\mathcal{E}_D}(\delta(S))$.

On the over side, λ is strictly increasing and then $\lambda(i_e) - \lambda(i_1) + 1 \geq \mu_{\overline{\mathcal{E}_D}}(e)$ then, with the assumption of Eq. (A.3), $\lambda(i_e) - \lambda(i_1) + 1 > \mu_{\delta(S)}(e)$. This a contradiction as the difference between indices of two elements having the same event type can be larger than the multiplicity of this event type. $\qquad\square$

Proposition 5.3 $abs(\mathcal{D}) \Subset S$, for all $S \in \mathcal{D}$

Proof of Proposition 5.3 We take the same notation as in the previous proof and we illustrate the results we have from now in the graphic below

This graphic illustrates the chronicles at hand. Each edge is an embedding between two multisets (from the smallest to the largest multiset). The dashed arrow is the focus of this proof.

We can also notice that λ_S derives from Definition 5.3. Thus, it is a least embedding. In addition, we know that $\theta = Id$ because the multiset are the same.

We also notice that all these chronicles are simples. This means that there is at most one temporal constraint between to events.

The objective is to prove that there exists an embedding τ that satisfies the temporal constraints.

Again ... let us proceed by contradiction ($abs(\mathcal{D}) \not\Subset S$). As we know that τ exists (Lemma 5.17), this means that we assume that for any τ, there exists at least a temporal constraint of $\overline{\mathcal{T}_D}$ that is not satisfied in S while applying the mapping τ.

More formally, we assume that there exists a temporal constraint $\left(e''_{i_c}, i_c\right)[l_c, u_c]$ $\left(e''_{j_c}, j_c\right) \in \overline{\mathcal{T}_{\mathcal{D}}}$ such that $d_c \notin [l_c, u_c]$ where d_c is the delay between $\tau(i_c)$ and $\tau(j_c)$ in S, i.e. $(e^S_{\tau(i_c)}, \tau(i_c))[d_c, d_c](e^S_{\tau(j_c)}, \tau(j_c))$.

We have that $\theta \circ \lambda_S \circ \tau$ is an embedding of $abs(\mathcal{D})$ in glb $\Lambda_{\mathcal{E}_{\mathcal{D}}}(\mathcal{D})$. By the uniqueness of this embedding (Lemma 5.16), $\theta \circ \lambda_S \circ \tau = \lambda_\epsilon$.

On the one side, $\left(e'_{\lambda_\epsilon(i_c)}, \lambda_\epsilon(i_c)\right)[l_c, u_c]\left(e'_{\lambda_\epsilon(j_c)}, \lambda_\epsilon(j_c)\right) \in \overline{\mathcal{T}_{\mathcal{D}}}$ is a temporal constraint in $\mathcal{T}_{\mathcal{D}}$. This means that $[l_c, u_c]$ is the temporal interval between the events with indices $\lambda_\epsilon(i_c)$ and $\lambda_\epsilon(j_c)$ (remind that in a simple chronicle there is at most on temporal constraint).

On the other side, $\left(e'_{\lambda_S(\tau(i_c))}, \lambda_S(\tau(i_c))\right)[d_c, d_c]\left(e'_{\lambda_S(\tau(j_c))}, \lambda_S(\tau(j_c))\right)$ is a temporal constraint of $\widetilde{\Lambda_{\mathcal{E}_{\mathcal{D}}}(\mathcal{T}_S)}$. According to Proposition 3.3, $\Lambda_{\mathcal{E}_{\mathcal{D}}}(\delta(S)) \preceq$ glb $\Lambda_{\mathcal{E}_{\mathcal{D}}}(\mathcal{D})$ and then $\left(e'_{\theta(\lambda_S(\tau(i_c)))}, \theta(\lambda_S(\tau(i_c)))\right)[l, u]\left(e'_{\theta(\lambda_S(\tau(j_c)))}, \theta(\lambda_S(\tau(j_c)))\right)$ $\trianglelefteq \left(e'_{\lambda_S(\tau(i_c))}, \lambda_S(\tau(i_c))\right)[d_c, d_c]\left(e'_{\lambda_S(\tau(j_c))}, \lambda_S(\tau(j_c))\right)$, ie $d_c \in [l, u]$, where $[l, u]$ is the temporal constraints between the event with indices $\theta(\lambda_S(\tau(i_c))) = \lambda_\epsilon(i_c)$ and $\theta(\lambda_S(\tau(j_c))) = \lambda_\epsilon(j_c)$. Then, $[l, u] = [l_c, u_c]$, but $d_c \notin [l_c, u_c]$. This is a contradiction that prove that our assumption was wrong and $abs(\mathcal{D}) \in S$.　　□

A.8　Proofs for Sect. 5.4.1

Proposition 5.4 If $\mathcal{C}$ and $\mathcal{C}'$ are two chronicles such that $\mathcal{C}' \preceq \mathcal{C}$ then $supp_{\mathcal{D}}(\mathcal{C}') \geq supp_{\mathcal{D}}(\mathcal{C})$.

Proof of Proposition 5.4 Since $\mathcal{C}' \preceq \mathcal{C}$, an easy consequence of Proposition 4.2 is $\{S \in \mathcal{D} \mid \mathcal{C} \in S\} \subseteq \{S \in \mathcal{D} \mid \mathcal{C}' \in S\}$. Then, apply Definition 5.6.　　□

A.9　Proofs for Sect. 5.4.4

Proposition 5.5 (φ, ψ) is a Galois connection between $(\widetilde{\mathcal{C}_{f\mu_{\mathcal{D}}}}, \preceq)$ and $(2^{Str(\mathcal{D})}, \supseteq$).

Proof of Proposition 5.5 First, it is easy to see that φ and ψ are well-defined: the image of a set of sequences S by ψ falls into $\widetilde{\mathcal{C}_{f\mu_{\mathcal{D}}}}$ and conversely by φ.

Then, the Proposition expresses the following property: $\varphi(\mathcal{C}) \supseteq S \iff \mathcal{C} \preceq \psi(S)$. The following proves this latter property.

Let $\mathcal{C} \in \widetilde{\mathcal{C}_f}$ and S be a set of sequences in $Str(\mathcal{D})$

$$S \subseteq \psi(\mathcal{C}) \iff \mathcal{C} \preceq \Lambda_{\mu_S}(\delta(S)), \forall S \in S \tag{A.4}$$

$$\Leftrightarrow \mathscr{C} \preceq \operatorname*{glb}_{S \in \mathcal{S}} \Lambda_{\mu_{\mathcal{D}}}(\delta(S)) \tag{A.5}$$

$$\Leftrightarrow \mathscr{C} \preceq \psi(\mathscr{C}) \tag{A.6}$$

(A.4) by definition of ψ. (A.5) All the chronicles at hand lies in the lattice of the chronicle that conforms exactly with $\mu_{\mathcal{D}}$. By definition of the least general bound in this lattice, the equivalence holds. Equation (A.6) by definition of φ. $\qquad\square$

Proposition 5.6 Let $\mathcal{D}$ be a dataset of sequences, then $(Str(\mathcal{D}), \widetilde{\mathcal{C}_{f\mu_{\mathcal{D}}}}, \Lambda_{\mu_{\mathcal{D}}} \circ \delta)$ is a pattern structure.

Proof of Proposition 5.6 Proposition 5.5 proves that there is a Galois connection between the powerset of stripped sequences and $\widetilde{\mathcal{C}_{f\mu_{\mathcal{D}}}}$. Moreover, $\widetilde{\mathcal{C}_{f\mu_{\mathcal{D}}}}$ is a complete semi-lattice and $Str(\mathcal{D})$ is finite. Then by definition, $(Str(\mathcal{D}), \widetilde{\mathcal{C}_{f\mu_{\mathcal{D}}}}, \Lambda_{\mu_{\mathcal{D}}} \circ \delta)$ is a pattern structure. $\qquad\square$

Lemma 5.19 The intent of a formal concept is a closed chronicle in the set of chronicles of all formal concepts.

Proof of Lemma 5.19 Let $\mathcal{D}$ be a dataset of temporal sequences and $(Str(\mathcal{D}), \widetilde{\mathcal{C}_{f\mathcal{E}_{\mathcal{D}}}}, \Lambda_{\mathcal{E}_{\mathcal{D}}} \circ \delta)$ a pattern structure (Proposition 5.6).

Let $(\mathscr{C}, \mathcal{S})$ be a formal concept, then according to Proposition 5.5, $\mathscr{C} = \psi(\mathcal{S})$ and $\mathcal{S} = \varphi(\mathscr{C})$.

Let $\mathscr{C}'$ be a chronicle such that $\mathscr{C} \prec \mathscr{C}'$. We denote $\mathcal{S}' = \varphi(\mathscr{C}')$. By Proposition 4.2, $\mathcal{S}' \subseteq \mathcal{S}$. Let us assume that $\mathcal{S}' = \mathcal{S}$.

Then $\mathscr{C}'$ occurs in all sequence $S \in \mathcal{S}$. It ensues $\mathscr{C}' \preceq \mathscr{C}$ because $\mathscr{C}$ is, by definition, the glb of the chronicles that occurs in $\mathcal{S}$. This is a contradiction and then, all greater chronicles have a strictly lower support. By Definition 5.7, $\mathscr{C}$ is closed. $\qquad\square$

Appendix B
Additional Content

B.1 Joint Intersection on the Set of Simple Chronicles

The class of simple chronicles is not closed under the intersection operation (Definition 3.4) that has been proposed in the previous section. An illustration is as follows. $\mathscr{C}_1 = (\{\!\{A, B, B\}\!\}, \{(A, 1)[3, 6](B, 2), (A, 1)[7, 9](B, 3)\})$ and $\mathscr{C}_2 = (\{\!\{A, B\}\!\}, \{(A, 1)[4, 8](B, 2)\})$ are two simple chronicles. By definition, $\mathscr{C}_1 \curlywedge \mathscr{C}_2 = (\{\!\{A, B\}\!\}, \{(A, 1)[3, 8](B, 2), (A, 1)[4, 9](B, 2)\})$ which is not a simple chronicle.

In traditional chronicle mining algorithms [27], the generalisation of sequences that is used differs from the generalisation proposed in our formal account (see Sect. 3.4). The following definitions introduce an adapted intersection operation for simple chronicles.

Definition B.1 (Simple Joint Intersection) Let $\mathscr{C} = (\mathcal{E} = \{\!\{e_1, \ldots, e_m\}\!\}, \mathcal{T})$ and $\mathscr{C}' = (\mathcal{E}'\{\!\{e'_1, \ldots, e'_{m'}\}\!\}, \mathcal{T}')$ be simple chronicles. Let Θ be the set of all multiset embeddings θ for $(\mathcal{E} \cap \mathcal{E}') \subseteq \mathcal{E}$ and Θ' the set of all multiset embeddings θ' for $(\mathcal{E} \cap \mathcal{E}') \subseteq \mathcal{E}'$. Define the **simple joint intersection** of $\mathscr{C}$ and $\mathscr{C}'$, denoted $\mathscr{C} \sqcap \mathscr{C}' = (\mathcal{E} \cap \mathcal{E}', \mathcal{T} \oplus \mathcal{T}')$, as

- $\mathcal{E} \cap \mathcal{E}' = \{\!\{e''_1, \ldots, e''_s\}\!\}$ is the multiset intersection of $\mathcal{E}$ and $\mathcal{E}'$,
- $\mathcal{T} \oplus \mathcal{T}' = \left\{ (e''_i, i)[l : u](e''_j, j) \mid \textsf{Proviso} \right\}$

  where $\textsf{Proviso}$ stands for

$$
\begin{aligned}
& 1 \leq i < j \leq s, \\
& \mathcal{S}_{ij} = \{(e''_i, \theta(i))[\lambda : \mu](e''_j, \theta(j)) \in \mathcal{T} \mid \theta \in \Theta\}, \\
& \mathcal{S}'_{ij} = \{(e''_i, \theta'(i))[\lambda : \mu](e''_j, \theta'(j)) \in \mathcal{T}' \mid \theta' \in \Theta'\}, \\
& \mathcal{S}_{ij} \neq \emptyset \text{ and } \mathcal{S}'_{ij} \neq \emptyset, \\
& u = \max\{\mu \mid (e''_i, \theta(i))[\lambda : \mu](e''_j, \theta(j)) \in \mathcal{S}_{ij} \cup \mathcal{S}'_{ij}\}, \\
& l = \min\{\lambda \mid (e''_i, \theta(i))[\lambda : \mu](e''_j, \theta(j)) \in \mathcal{S}_{ij} \cup \mathcal{S}'_{ij}\}.
\end{aligned}
$$

© The Author(s), under exclusive license to Springer Nature Switzerland AG 2023
T. Guyet, P. Besnard, *Chronicles: Formalization of a Temporal Model*,
SpringerBriefs in Computer Science, https://doi.org/10.1007/978-3-031-33693-5

Remark That max and min exist in Definition B.1 is a consequence of the following observation: $\mathcal{E}$ and $\mathcal{E}'$ are finite and there is at most one temporal constraint between each pair of events, so $\mathcal{T}$ and $\mathcal{T}'$ are also finite. The condition that $\mathcal{S}_{ij}$ and $\mathcal{S}'_{ij}$ be non-empty means that if $\mathcal{S}_{ij}$ or $\mathcal{S}'_{ij}$ is empty, then the resulting simple chronicle will have not any constraints between events i and j.

Remark $\mathscr{C} \sqcap \mathscr{C}'$ is a simple chronicle. By definition, $\mathcal{T} \oplus \mathcal{T}'$ contains at most one constraint between each pair of events of $\mathcal{E} \cap \mathcal{E}'$.

Intuitively, the elements of $\mathcal{T} \oplus \mathcal{T}'$ in Definition B.1 are computed as the convex-hull of the intervals between same event types. A graphical illustration for Definition B.1 is provided by Fig. B.1. This Figure compares the intersection obtained using each of the intersection operators. Both intersection share the multiset $\{\!\{A, B\}\!\}$, but differ in the temporal constraints. In the joint intersection, there are two temporal constraints due to the existence of two different embeddings of $\mathcal{E} \cap \mathcal{E}'$ in the chronicle $\mathscr{C}'$. In the simple joint intersection, the boundaries $l_{\ominus'} = 2$ and $u_{\ominus'} = 10$ are obtained by computing the convex-hull of the intervals between event types A and B, i.e., between the intervals $[2 : 3]$ and $[5 : 10]$. The final temporal constraint is the convex-hull of the intervals obtained by the boundaries of the two chronicles $\mathscr{C}$ and $\mathscr{C}'$, resp. $[3 : 7]$ and $[2 : 10]$, i.e., the interval $[2 : 10]$.

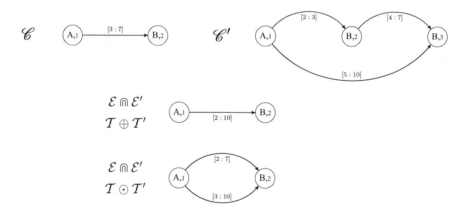

Fig. B.1 On top: two simple chronicles. In the middle: intersection of two simple chronicles using the adapted joint intersection. At the bottom: intersection of two simple chronicles using the original joint intersection

References

1. James F. Allen and George Ferguson. Actions and events in interval temporal logic. *Journal of Logic and Computation*, 4(5):531–579, 1994.
2. Miguel R Álvarez, Paulo Félix, Purificación Cariñena, and Abraham Otero. A data mining algorithm for inducing temporal constraint networks. In *Proc. of the Int. Conf. on Information Processing and Management of Uncertainty in Knowledge-Based Systems*, pages 300–309. Springer, Jun 28-Jul 2, 2010.
3. Darko Anicic, Paul Fodor, Sebastian Rudolph, and Nenad Stojanovic. EP-SPARQL: A unified language for event processing and stream reasoning. In Sadagopan Srinivasan, Krithi Ramamritham, Arun Kumar, M. P. Ravindra, Elisa Bertino, and Ravi Kumar, editors, *Proc. of the 20th Int. Conf. on World Wide Web (WWW 2011)*, pages 635–644. ACM, March 28 - April 1, 2011.
4. Darko Anicic, Sebastian Rudolph, Paul Fodor, and Nenad Stojanovic. Stream reasoning and complex event processing in ETALIS. *Semantic Web*, 3(4):397–407, 2012.
5. Alexander Artikis, Marek J. Sergot, and Georgios Paliouras. An event calculus for event recognition. *IEEE Transactions on Knowledge and Data Engineering*, 27(4):895–908, 2015.
6. Alexander Artikis, Anastasios Skarlatidis, François Portet, and Georgios Paliouras. Logic-based event recognition. *The Knowledge Engineering Review*, 27(4):469–506, 2012.
7. Johanne Bakalara, Thomas Guyet, Olivier Dameron, André Happe, and Emmanuel Oger. An extension of chronicles temporal model with taxonomies-application to epidemiological studies. In *Proc. of the Int. Conf. on Informatics and Assistive Technologies for Health-Care, Medical Support and Wellbeing (HEALTHINFO 2021)*, pages 133–142. SCITEPRESS, October 3–7, 2021.
8. José L Balcázar and Gemma C Garriga. Horn axiomatizations for sequential data. *Theoretical Computer Science*, 371(3):247–264, 2007.
9. Philippe Besnard and Guyet Thomas. Semantics of negative sequential patterns. In *Proc. of the European Conf. on Artificial Intelligence (ECAI 2020)*, pages 1009–1015. IOS Press, 2020.
10. Wayne D. Blizard. Multiset theory. *Notre Dame Journal of Formal Logic*, 30(1):36–66, 1989.
11. Patricia Bouyer. Model-checking timed temporal logics. *Electronic notes in theoretical computer science*, 231:323–341, 2009.
12. Rajkumar Buyya and Amir Vahid Dastjerdi. *Internet of Things: Principles and Paradigms*. Elsevier, 2016.
13. Pedro Cabalar, Ramón P. Otero, and Silvia Gómez Pose. Temporal constraint networks in action. In Werner Horn, editor, *Proc. of the 14th European Conf. on Artificial Intelligence (ECAI 2000)*, pages 543–547. IOS Press, August 20–25, 2000.

14. Qiushi Cao, Ahmed Samet, Cecilia Zanni-Merk, François de Bertrand de Beuvron, and Christoph Reich. Combining chronicle mining and semantics for predictive maintenance in manufacturing processes. *Semantic Web*, 11(6):927–948, 2020.

15. Patrice Carle, Christine Choppy, Romain Kervarc, and Ariane Piel. Safety of unmanned aircraft systems facing multiple breakdowns. In Christine Choppy and Jun Sun, editors, *Proc. of the 1st French Singaporean Workshop on Formal Methods and Applications (FSFMA 2013)*, pages 86–91. OASICS 31, Schloss Dagstuhl - Leibniz-Zentrum für Informatik, July 15–16, 2013.

16. Guy Carrault, Marie-Odile Cordier, René Quiniou, and Feng Wang. Temporal abstraction and inductive logic programming for arrhythmia recognition from electrocardiograms. *Artificial Intelligence in Medicine*, 28(3):231–263, 2003.

17. Gemma Casas-Garriga. Summarizing sequential data with closed partial orders. In *Proc. of the SIAM Int. Conf. on Data Mining (SDM)*, pages 380–391. SIAM, March 17–21, 2005.

18. Saoussen Cheikhrouhou, Slim Kallel, Nawal Guermouche, and Mohamed Jmaiel. Toward a time-centric modeling of business processes in BPMN 2.0. In *Proc of IntConf on Information Integration and Web-based Applications & Services*, pages 154–163, December 2–4, 2013.

19. Carlo Combi, Barbara Oliboni, and Francesca Zerbato. Modeling and handling duration constraints in bpmn 2.0. In *Proc. of the 32st ACM Symposium on Applied Computing (SAC 2017)*, pages 727–734. Association for Computing Machinery, April 3–7, 2017.

20. Carlo Combi, Francesca Galetto, Hirenkumar Chandrakant Nakawala, Giuseppe Pozzi, and Francesca Zerbato. Enriching surgical process models by BPMN extensions for temporal durations. In *Proc. of the 36th Annual ACM Symposium on Applied Computing*, pages 586–593 (2021).

21. Damien Cram, Benoit Mathern, and Alain Mille. A complete chronicle discovery approach: Application to activity analysis. *Expert Systems*, 29(4):321–346, 2012.

22. Yann Dauxais. *Discriminant Chronicle Mining*. PhD thesis, Univ. Rennes, France, April 2018.

23. Yann Dauxais and Thomas Guyet. Generalized chronicles for temporal sequence classification. In *Proc. of Workshop on Advanced Analytics and Learning on Temporal Data (AALTD@ECML/PKDD)*, pages 30–45. Springer International, September 2020.

24. Yann Dauxais, Thomas Guyet, David Gross-Amblard, and André Happe. Discriminant chronicles mining. In Annette ten Teije, Christian Popow, John H. Holmes, and Lucia Sacchi, editors, *Proc. of the 16th Conf. on Artificial Intelligence in Medicine in Europe (AIME 2017)*, volume 10259 of *Lecture Notes in Computer Science*, pages 234–244. Springer, June 21–24, 2017.

25. Rina Dechter, Itay Meiri, and Judea Pearl. Temporal constraint networks. *Artificial Intelligence*, 49(1–3):61–95, 1991.

26. Stéphane Demri, Valentin Goranko, and Martin Lange. *Temporal Logics in Computer Science: Finite-State Systems*, volume 58 of *Cambridge Tracts in Theoretical Computer Science*. Cambridge University Press, 2016.

27. Christophe Dousson and Thang Vu Duong. Discovering chronicles with numerical time constraints from alarm logs for monitoring dynamic systems. In Thomas Dean, editor, *Proc. of the 16th Int. Joint Conf. on Artificial Intelligence (IJCAI 1999)*, pages 620–626. Morgan Kaufmann, July 31 - August 6, 1999.

28. Christophe Dousson, Paul Gaborit, and Malik Ghallab. Situation recognition: Representation and algorithms. In Ruzena Bajcsy, editor, *Proc. of the 13th Int. Joint Conf. on Artificial Intelligence (IJCAI 1993)*, pages 166–172. Morgan Kaufmann, August 28 - September 3, 1993.

29. Christophe Dousson and Pierre Le Maigat. Chronicle recognition improvement using temporal focusing and hierarchization. In Manuela M. Veloso, editor, *Proc. of the 20th Int. Joint Conf. on Artificial Intelligence (IJCAI 2007)*, pages 324–329, January 6–12, 2007.

30. Anton Dries and Siegfried Nijssen. Mining patterns in networks using homomorphism. In *Proc. of the 12th SIAM Int. Conf. on Data Mining (SDM 2012)*, pages 260–271. SIAM/Omnipress, April 26–28, 2012.

31. Denis Gagne and André Trudel. Time-BPMN. In *Proc. of the IEEE Conf. on Commerce and Enterprise Computing (CEC 2009)*, pages 361–367. IEEE, July 20–23, 2009.

32. Bernhard Ganter and Sergei O. Kuznetsov. Pattern structures and their projections. In Harry S. Delugach and Gerd Stumme, editors, *Proc. of the 9th Int. Conf. on Conceptual Structures (ICCS 2001)*, volume 2120 of *Lectures Notes in Computer Science*, pages 129–142. Springer, July 30 - August 3, 2001.

33. Bernhard Ganter and Rudolf Wille. *Formal Concept Analysis: Mathematical Foundations*. Springer, 1999.

34. Paul Gastin and Denis Oddoux. Fast LTL to Büchi automata translation. In Gérard Berry, Hubert Comon, and Alain Finkel, editors, *Proc. of the 13th Int. Conf. on Computer Aided Verification (CAV 2001)*, volume 2102 of *Lecture Notes in Computer Science*, pages 53–65. Springer, July 18–22, 2001.

35. Clément Gautrais, René Quiniou, Peggy Cellier, Thomas Guyet, and Alexandre Termier. Purchase signatures of retail customers. In Jinho Kim, Kyuseok Shim, Longbing Cao, Jae-Gil Lee, Xuemin Lin, and Yang-Sae Moon, editors, *Proc. of the 21st Pacific-Asia Conf. on Knowledge Discovery and Data Mining (PAKDD 2017)*, volume 10234 of *Lecture Notes in Computer Science*, pages 110–121. Springer, May 23–26, 2017.

36. Malik Ghallab. On chronicles: Representation, on-line recognition and learning. In Luigia Carlucci Aiello, Jon Doyle, and Stuart C. Shapiro, editors, *Proc. of the 5th Int. Conf. on Principles of Knowledge Representation and Reasoning (KR 1996)*, pages 597–606. Morgan Kaufmann, November 5–8, 1996.

37. Fosca Giannotti, Mirco Nanni, Dino Pedreschi, and Fabio Pinelli. Mining sequences with temporal annotations. In Hisham Haddad, editor, *Proc. of the 21st ACM Symposium on Applied Computing (SAC 2006)*, pages 593–597, April 23–27, 2006.

38. Syed Gillani, Antoine Zimmermann, Gauthier Picard, and Frédérique Laforest. A query language for semantic complex event processing: Syntax, semantics and implementation. *Semantic Web*, 10(1):53–93, 2019.

39. Alejandro Grez, Cristian Riveros, and Martín Ugarte. Foundations of complex event processing. *CoRR*, abs/1709.05369, 2017.

40. Thomas Guyet, Philippe Besnard, Ahmed Samet, Nasreddine Ben Salha, and Nicolas Lachiche. Énumération des occurrences d'une chronique. In Antoine Cornuéjols and Etienne Cuvelier, editors, *Extraction et Gestion des Connaissances, EGC 2020, Brussels, Belgium, January 27–31, 2020*, volume E-36, pages 253–260. Éditions RNTI, 2020.

41. Thomas Guyet and René Quiniou. NegPSpan: Efficient extraction of negative sequential patterns with embedding constraints. *CoRR* abs/1804.01256, 2018.

42. Jiawei Han, Micheline Kamber, and Jian Pei. Mining frequent patterns, associations, and correlations: Basic concepts and methods. In Jiawei Han, Micheline Kamber, and Jian Pei, editors, *Data Mining (Third Edition)*, The Morgan Kaufmann Series in Data Management Systems, chapter 6, pages 243–278. Morgan Kaufmann, Boston, third edition edition, 2012.

43. Pavol Hell and Jaroslav Nešetřil. *Graphs and Homomorphisms*, volume 28 of *Oxford Lecture Series in Mathematics and Its Applications*. Oxford University Press, 2004.

44. Frank Höppner. Learning dependencies in multivariate time series. In *Proc. of the Workshop on Knowledge Discovery in Spatio-Temporal Data (Workshop at ECAI 2002)*, pages 25–31, July 21–26, 2002.

45. Chuntao Jiang, Frans Coenen, and Michele Zito. A survey of frequent subgraph mining algorithms. *Knowledge Engineering Review*, 28(1):75–105, 2013.

46. Robert Kowalski and Marek J. Sergot. A logic-based calculus of events. In Joachim W. Schmidt and Costantino Thanos, editors, *Foundations of Knowledge Base Management*, Topics in Information Systems, pages 23–55. Springer, 1989.

47. Sergei O. Kuznetsov. Pattern structures for analyzing complex data. In Hiroshi Sakai, Mihir K. Chakraborty, Aboul Ella Hassanien, Dominik Slezak, and William Zhu, editors, *Proc. of the 12th Int. Conf. on Rough Sets, Fuzzy Sets, Data Mining, and Granular Computing (RSFDGrC 2009)*, volume 5908 of *Lecture Notes in Computer Science*, pages 33–44. Springer, December 15–18, 2009.

48. Sergei O. Kuznetsov and Mikhail V. Samokhin. Learning closed sets of labeled graphs for chemical applications. In Stefan Kramer and Bernhard Pfahringer, editors, *Proc. of the 15th Int.*

Conf. on Inductive Logic Programming (ILP 2005), volume 3625 of *Lecture Notes in Computer Science*, pages 190–208. Springer, August 10–13, 2005.

49. Xavier Le Guillou, Marie-Odile Cordier, Sophie Robin, Laurence Rozé, et al. Chronicles for on-line diagnosis of distributed systems. In *Proc. of the European Conf. on Artificial Intelligence (ECAI 2008)*, pages 194–198. IOS Press, 2008.

50. T. Y. Cliff Leung and Richard R. Muntz. Temporal query processing and optimization in multiprocessor database machines. In Li-Yan Yuan, editor, *Proc. of the 18th Int. Conf. on Very Large Data Bases (VLDB 1992)*, pages 383–394. Morgan Kaufmann, August 23–27, 1992.

51. Hector Levesque, Raymond Reiter, Yves Lespérance, Fangzhen Lin, and Richard Scherl. Golog: A logic programming language for dynamic domains. *The Journal of Logic Programming*, 31(1):59–83, 1997.

52. Carsten Lutz, Frank Wolter, and Michael Zakharyaschev. Temporal description logics: A survey. In Stéphane Demri and Christian S. Jensen, editors, *Proc. of the 15th Int. Symposium on Temporal Representation and Reasoning (TIME 2008)*, pages 3–14. IEEE, June 16–18, 2008.

53. Nizar R. Mabroukeh and Christie I. Ezeife. A taxonomy of sequential pattern mining algorithms. *ACM Computing Surveys*, 43(1):3:1–3:41, 2010.

54. Heikki Mannila, Hannu Toivonen, and A. Inkeri Verkamo. Discovery of frequent episodes in event sequences. *Data Mining and Knowledge Discovery*, 1(3):259–289, 1997.

55. John McCarthy. Situations, actions, and causal laws. Technical report, Dept of Computer Science, Stanford University, CA, USA, 1963.

56. Fabian Mörchen. Unsupervised pattern mining from symbolic temporal data. *SIGKDD Explorations: Newsletter of the Special Interest Group on Knowledge Discovery & Data Mining*, 9(1):41–55, 2007.

57. Erik T. Mueller. Event calculus. In Frank van Harmelen, Vladimir Lifschitz, and Bruce W. Porter, editors, *Handbook of Knowledge Representation*, volume 3 of *Foundations of Artificial Intelligence*, pages 671–708. Elsevier, 2008.

58. Erik T. Mueller. *Commonsense Reasoning: An Event Calculus Based Approach*. Morgan Kaufmann, 2nd edition, 2015.

59. Cristina Nica, Agnès Braud, Xavier Dolques, Marianne Huchard, and Florence Le Ber. Exploring temporal data using relational concept analysis: An application to hydroecology. In Marianne Huchard and Sergei O. Kuznetsov, editors, *Proc. of 13th Int. Conf. on Concept Lattices and Their Applications (CLA 2016)*, volume 1624 of *CEUR Workshop Proceedings*, pages 299–311. CEUR-WS.org, July 18–22, 2016.

60. Amir Pnueli. The temporal logic of programs. In *Proc. of the 18th Annual Symposium on Foundations of Computer Science (FOCS 1977)*, pages 46–57. IEEE, October 31 - November 1, 1977.

61. Raymond Reiter. *Knowledge in Action: Logical Foundations for Specifying and Implementing Dynamical Systems*. MIT Press, 2001.

62. Aurelian Rădoacă. Properties of multisets compared to sets. Technical report, Dept of Computer Science, West University of Timişoara, Romania, 2018.

63. Alexandre Sahuguède, Euriell Le Corronc, and Marie-Véronique Le Lann. An ordered chronicle discovery algorithm. In *Proc. of the 3nd ECML/PKDD Workshop on Advanced Analytics and Learning on Temporal Data (AALTD)*, Sep 10–14, 2018.

64. Eddie Schwalb and Rina Dechter. Processing disjunctions in temporal constraint networks. *Artificial Intelligence*, 93(1):29–61, 1997.

65. Eddie Schwalb and Lluís Vila. Temporal constraints: A survey. *Constraints*, 3(2):129–149, 1998.

66. Murray Shanahan. The event calculus explained. In Michael J. Wooldridge and Manuela Veloso, editors, *Artificial Intelligence Today*, volume 1600 of *Lecture Notes in Computer Science*, pages 409–430. Springer, 1999.

67. Richard T. Snodgrass. The temporal query language TQuel. *ACM Transactions on Database Systems (TODS)*, 12(2):247–298, 1987.

68. Richard T. Snodgrass, editor. *The TSQL2 Temporal Query Language*, volume 330 of *The Springer International Series in Engineering and Computer Science*. Springer, 1995.

69. Gerd Stumme. Efficient data mining based on formal concept analysis. In A. Hameurlain, R. Cicchetti, and R. Traunmüller, editors, *Proc. of the Int. Conf. on Database and Expert Systems Applications (DEXA 2002)*, volume 2453 of *Lectures Notes in Computer Science*, pages 534–546. Springer, September 2–6, 2002.

70. Jonas Tappolet and Abraham Bernstein. Applied temporal RDF: Efficient temporal querying of RDF data with SPARQL. In Lora Aroyo, Paolo Traverso, Fabio Ciravegna, Philipp Cimiano, Tom Heath, Eero Hyvönen, Riichiro Mizoguchi, Eyal Oren, Marta Sabou, and Elena Paslaru Bontas Simperl, editors, *Proc. of the 6th European Semantic Web Conference (ESWC 2009)*, volume 5554 of *Lecture Notes in Computer Science*, pages 308–322. Springer, May 31 - June 4, 2009.

71. Katerina Tsesmeli, Manel Boumghar, Thomas Guyet, René Quiniou, and Laurent Pierre. Fouille de motifs temporels négatifs. In Mustapha Lebbah, Christine Largeron, and Hanane Azzag, editors, *Actes des 18èmes Journées Francophones Extraction et Gestion des Connaissances (EGC 2018)*, pages 263–268. RNTI, 22–26 janvier, 2018.

72. Dogan Ulus and Oded Maler. Specifying timed patterns using temporal logic. In Maria Prandini and Jyotirmoy V. Deshmukh, editors, *Proc. of the 21st Int. Conf. on Hybrid Systems: Computation and Control (HSCC 2018)*, pages 167–176. ACM, April 11–13, 2018.

73. Kristof Van Belleghem, Marc Denecker, and Danny De Schreye. Combining situation calculus and event calculus. In Leon Sterling, editor, *Proc. of the 12th Int. Conf. on Logic Programming*, pages 83–97. MIT Press, June 13–16, 1995.

74. Natalia Vanetik, Ehud Gudes, and Solomon Eyal Shimony. Computing frequent graph patterns from semistructured data. In *Proc. of the Int. Conf. on Data Mining (ICDM 2002)*, pages 458–465. IEEE Computer Society, Decembre, 9–12 2002.

75. Show-Jane Yen and Yue-Shi Lee. Mining non-redundant time-gap sequential patterns. *Applied Intelligence*, 39(4):727–738, 2013.

Index

T. Guyet, P. Besnard, *Chronicles: Formalization of a Temporal Model*,
SpringerBriefs in Computer Science, https://doi.org/10.1007/978-3-031-33693-5

Printed in the United States
by Baker & Taylor Publisher Services